Studies in STRUCTURE

JOAN M. HOLLAND

Formerly Senior Lecturer in Mathematics,
Bishop Otter College, Chichester

MACMILLAN

First published 1972

Published by
MACMILLAN PRESS LTD
Basingstoke and London
The Macmillan Company of Australia Pty Ltd Melbourne
The Macmillan Company of Canada Ltd Toronto
St Martin's Press Inc New York
Companies and representatives
throughout the world

SBN 333 11345 4

Printed in Great Britain by
Robert MacLehose & Co. Ltd.
The University Press, Glasgow

PREFACE

This book began as a record of investigations made in pursuit of the author's personal mathematical interests. Such material is to be found mainly in Chapters 5, 7–9 (the latter half of each), 10, and 11. The rest is a modest attempt to fit the investigations into their proper context of abstract algebraic systems and processes already well established and clearly formulated in a number of authoritative textbooks.

For many students, including the author, the way to an appreciation of abstract mathematics of this kind is made easier by preliminary study of a variety of particular cases which have interesting structural features. Hence the investigations mentioned above with others of a more conventional nature have been used freely throughout the book to serve several distinct purposes. First, they have been used to prepare readers, by informal discussion, for precise definition in later sections, giving them a chance of making their own abstractions ahead of schedule. Then, when an abstract system such as a *group* or a special term such as *homomorphism* has been formally defined, the investigations have been used to assist understanding of the definition both in reference back to previous cases and as a source of fresh illustrations. Another purpose is to develop the reader's ability not only to recognise a general structure such as *group*, *ring*, or *field* but also to identify and name a particular group which may be encountered in first one and then another of a variety of situations. This ability contributes to a proper appreciation of the important principle of *isomorphism*.

At appropriate points in the text, readers are invited to make their own investigations of particular structures. These exercises are designed to point the way towards future developments, to evoke relevant abstract ideas, and to provide opportunities for practising techniques already introduced.

Although primarily intended for students following mathematical courses in Sixth Forms and Colleges of Education the book will be of interest to non-specialist students and to general readers with mathematical tastes. Most of the content demands no more than O level mathematics as a fund of previous knowledge. For science students especially, some of the topics treated and the methods of approach may be in harmony with their main course. It is possible also that lecturers in Colleges of Education and Sixth Form teachers may find this blend of abstract and concrete algebraic material of some use when they are preparing schemes of work.

I should like to express my thanks to my colleagues at Bishop Otter College, particularly Lenon Beeson, Jean Dilnot, and Tim Brook for their valuable comments and help in providing material; also to the students who assisted me greatly by their response to some of the topics. Grateful acknowledgements are also due to Professor Walter Ledermann of the University of Sussex for his advice and encouragement at an early stage, and to the late Mr. A. J. Moakes for his valuable criticism and helpful suggestions; also to Mr. Wilfrid Wilson who kindly offered to read the section on change-ringing of bells and suggested several amendments. I am grateful also to Peggy Sidwell, Barbara Lallemand, and Celia Crowley for much patience in typing the manuscript.

CONTENTS

MODULAR ARITHMETIC

1.1 Odd and even numbers

The arithmetic of odd and even numbers is well-known. If an odd number is combined with an even number by addition, the result is odd; if combined by multiplication the result is even. Tables can be constructed for all possible combinations:

Addition   Multiplication

$+$	e	o
e	e	o
o	o	e

$\times$	e	o
e	e	e
o	e	o

To speak of odd and even numbers implies a classification of the set of natural numbers together with zero into two sub-sets:

even	0	2	4	6	. . . .
odd	1	3	5	7	. . . .

1.2 Modulus 3

Even numbers divided by 2 give a remainder 0, odd numbers a remainder 1, so that the sub-sets can be named class 0 and class 1 respectively. This idea can be extended, using 3 as a divisor. Three sub-sets appear:

divisible by 3	0, 3, 6, 9 . . . class 0,
remainder 1	1, 4, 7, 10 . . . class 1,
remainder 2	2, 5, 8, 11 . . . class 2.

The three classes consist of numbers of the form $3n$, $3n+1$, $3n+2$ respectively. Tables can be constructed as before:

Addition

+	0	1	2
0	0	1	2
1	1	2	0
2	2	0	1

Multiplication

×	0	1	2
0	0	0	0
1	0	1	2
2	0	2	1

As the processes are not so familiar as those of odd and even numbers these tables should perhaps be explained in greater detail.

According to the addition table,

$$2+1=0.$$

This statement sums up an infinite number of true statements in which any representative of a class may be taken as typical of its class.

e.g. 5 (in class 2)$+$ 7 (in class 1)$=$ 12 (in class 0)

50 (in class 2)$+$67 (in class 1)$=$117 (in class 0).

In the same way the statement

$$2\times2=1 \tag{1.1}$$

taken from the multiplication table, is a general one covering many particular cases amongst which may be found, for example,

8 (in class 2)$\times$11 (in class 2)$=$a number in class 1.

This is borne out by the fact that

$$8\times11=(3\times29)+1.$$

The generality of equation (1.1) is also obvious because

$$(3m+2)(3n+2)=9mn+3(2m+2n)+3+1$$
$$=(a\ multiple\ of\ 3)+1.$$

1.3 Looking for patterns

Let us look more closely at these two tables, to see if any clear pattern emerges. First the addition table. This shows some regularity: each

2

element 0, 1, 2, appears once in every row and column; the elements run in diagonal lines:

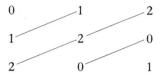

The pattern, if any, in the multiplication table is by no means obvious so let us try a new divisor or *modulus* as it is usually called. Modulus five breaks the natural numbers into 5 sets with remainders, or 'residues', of 0, 1, 2, 3, or 4.

Residue

0		0	5	10	15	.	.	.	class 0
1		1	6	11	16	.	.	.	class 1
2		2	7	12	17	.	.	.	class 2
3		3	8	13	18	.	.	.	class 3
4		4	9	14	19	.	.	.	class 4

Two members of the same class are said to be congruent to each other and the sign for congruency is $\equiv$. Thus $17 \equiv 2$ (mod 5). Where there is no danger of ambiguity such a statement is often written $17 = 2$ (mod 5). It may be taken as obvious that if $a \equiv b$, then $(a-b) \equiv 0$. Tables can be constructed as shown:

+	0	1	2	3	4
0	0	1	2	3	4
1	1	2	3	4	0
2	2	3	4	0	1
3	3	4	0	1	2
4	4	0	1	2	3

×	0	1	2	3	4
0	0	0	0	0	0
1	0	1	2	3	4
2	0	2	4	1	3
3	0	3	1	4	2
4	0	4	3	2	1

Exercise

1.1 Write out the addition tables for modulus 4 and modulus 6.

The addition table for modulus 5, as might have been expected, shows the same cyclic pattern as for modulus 3: each row is like the row before it, except that the elements have been displaced one step to the left. Now examine the multiplication table. Disregarding zeros, there remains a four-by-four array of elements:

×	1	2	3	4
1	1	2	3	4
2	2	4	1	3
3	3	1	4	2
4	4	3	2	1

There is some regularity: each element appears once in every row and column but the arrangement in rows is irregular and could not easily be foreseen. A slight rearrangement however will improve matters. Change the order of the multipliers outside the table from 1, 2, 3, 4 to 1, 2, 4, 3, and the table would then state the same facts but would look like this:

×	1	2	4	3
1	1	2	4	3
2	2	4	3	1
4	4	3	1	2
3	3	1	2	4

We are now back on familiar ground. The same diagonal pattern has appeared in multiplication as in addition and there is the interesting possibility of finding a direct link between the two processes that is worth investigating.

1.4 Isomorphism

On the one hand, we have a cyclic relationship between 4 elements obtained from using modulus 5 for multiplication. On the other hand, addition, using modulus 4, leads to exactly the same pattern:

+	0	1	2	3
0	0	1	2	3
1	1	2	3	0
2	2	3	0	1
3	3	0	1	2

This striking resemblance is an example of an extremely important and far-reaching principle called *isomorphism*, a word of Greek derivation meaning 'having the same shape'. It is a term which will later require precise definition. Its nature can be most easily grasped at first by considering particular instances and the tables above make a good starting point.

Because the four elements 1, 2, 4, 3 for multiplication (modulus 5) form a system isomorphic with the four elements 0, 1, 2, 3 for addition (modulus 4), any manipulation of elements in the one system is faithfully reflected in the other. Let us consider a few examples of this property and begin by setting out the correspondences between the elements of the two systems.

Multiplication (mod 5)		*Addition (mod 4)*
1	⟷	0
2	⟷	1
4	⟷	2
3	⟷	3

Examples

1. $(4 \times 3)\,(\bmod 5)$ ⟷ $(2+3)\,(\bmod 4),$
   i.e. $12\ (\bmod 5)$ ⟷ $5\ (\bmod 4),$
   i.e. $2\ (\bmod 5)$ ⟷ $1\ (\bmod 4).$

Now confirm that this last line is true by looking at the second correspondence above.

5

2. *Muliplication (mod 5)*　　　　　　　　　　*Addition (mod 4)*

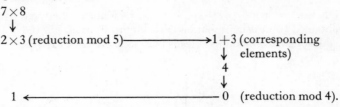

7×8
$\downarrow$
2×3 (reduction mod 5)──────→$1+3$ (corresponding
　　　　　　　　　　　　　　　　　　　$\downarrow$　　elements)
　　　　　　　　　　　　　　　　　　　4
　　　　　　　　　　　　　　　　　　　$\downarrow$
　　1 ←──────────────── 0 　(reduction mod 4).

This is confirmed by the fact that $7 \times 8 = 56 = 1$ (mod 5). The isomorphism allows us to pass at will from one system to the other: each
element has its 'image' in the other system; multiplication in one
system and addition in the other are also mutual images.

3.　Since multiplication in one system corresponds to addition in the
other:

$$4 \times 4 \times 4 \text{ (mod 5)} \longleftrightarrow 2+2+2 \text{ (mod 4)}$$
$$4^3 \quad \text{(mod 5)} \longleftrightarrow 3 \times 2 \quad \text{(mod 4)},$$

and in general repeated multiplication in one system corresponds to
repeated addition in the other.

$$4^n \quad \text{(mod 5)} \longleftrightarrow n \times 2 \quad \text{(mod 4)}.$$

By using this property we can solve such problems as examples 4 and 5
below.

4.　Show that $2^6 + 1$ is divisible by 5.

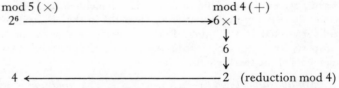

　　mod 5 $(\times)$　　　　　　　　　　mod 4 $(+)$
　　2^6 ──────────────→6×1
　　　　　　　　　　　　　　　　　　$\downarrow$
　　　　　　　　　　　　　　　　　　6
　　　　　　　　　　　　　　　　　　$\downarrow$
　　4 ←──────────────── 2 　(reduction mod 4)

which shows that $2^6 = 4$. Hence in the mod 5 system $2^6 + 1$ must be
divisible by 5. This calculation can readily be checked by ordinary
arithmetic but the next example which is easy to solve by the principle
of isomorphism would be extremely tedious by ordinary arithmetic.

5.　Show that $3^{142} + 1$ is divisible by 5.

　　Mod 5 $(\times)$　　　　　　　　　　Mod 4 $(+)$
　　3^{142} ──────────────→142×3
　　　　　　　　　　　　　　　　　　$\downarrow$
　　　　　　　　　　　　　　　　　　426
　　　　　　　　　　　　　　　　　　$\downarrow$
　　4 ←──────────────── 2

The divisibility of $(3^{142} + 1)$ by 5 follows immediately.

1.5 Another example of isomorphism

It may have been noticed that the table for addition of odd and even numbers is isomorphic with the table commonly used for the rule of signs for multiplication of numbers.

$+$	e	o
e	e	o
o	o	e

$\times$	$+$	$-$
$+$	$+$	$-$
$-$	$-$	$+$

We could word the rule for addition of odd and even numbers as follows:
'Like numbers give evens, unlike numbers give odds'
which is an exact parallel to the rule for multiplication:
'Like signs give plus, unlike signs give minus'.

1.6 Extension of the isomorphic principle

The correspondences between the two sets of four elements seem to have yielded useful and consistent results, based on an isomorphism, but too much importance should not be attached to a single instance of such a general principle. More investigation is required. We can at once ask some pertinent questions:

1. Does it work for any other modulus?
2. Does it work for all moduli?
3. In the relationship already examined would the isomorphism hold for another way of matching the elements, still retaining the same cyclic order, e.g. 0, 1, 2, 3 (mod 4 +)$\longleftrightarrow$2, 4, 3, 1 (mod 5 $\times$)?

Let us examine question 3 first. The tables would look like this:

Mod 4

$+$	0	1	2	3
0	0	1	2	3
1	1	2	3	0
2	2	3	0	1
3	3	0	1	2

Mod 5

$\times$	2	4	3	1
2	4	3	1	2
4	3	1	2	4
3	1	2	4	3
1	2	4	3	1

So far so good: the tables look alike and both contain nothing but true statements. Now see whether an operation in one system is always perfectly reflected in the other. A single example is enough to show that this is not so:

$$1+2=3 \ (\text{mod } 4)$$
$$4\times3=2 \ (\text{mod } 5).$$

Although $1+2$ (mod 4) appears to correspond with 4×3 (mod 5), 3 (mod 4) does not appear to correspond with 2 (mod 5). Why does the apparent isomorphism not yield true results in this case? The reader might like to think this out for himself so the question has been put again in exercise 1.6, and the solution will be found in section 1.11.

Exercises

1.2 Use the correspondences in 1.6 (a) below to find a counterpart in arithmetic modulo 5 for each of the following statements in arithmetic modulo 4:

(a) $3+0=3$,
(b) $2+3+1=2$,
(c) $3\times5 \quad =3$.

1.3 Show that 4^n (mod 5)$=1$ or 4 according as n is even or odd.

1.4 Determine which of the following numbers are divisible by 5:

$$3^5-2, \ 2^{154}+3, \ 4^{1000}-1.$$

1.5 If 3^x+2 is divisible by 5, what are the possible values for x?

1.6 Consider the correspondences below:

	mod 4+		mod 5×		mod 4+		mod 5×
	0	⟷	1		0	⟷	2
(a)	1	⟷	2	(b)	1	⟷	4
	2	⟷	4		2	⟷	3
	3	⟷	3		3	⟷	1

(a) is an isomorphism and gives reliable correspondences between operations in the two sets of elements: (b) is not an isomorphism although the tables for addition modulo 4 and multiplication modulo 5 look alike when set out in the order suggested for the elements. What makes the important difference between (a) and (b)?

1.7 Write out in full the multiplication table modulo 6.

1.7 The use of moduli in everyday life

Let us break off from isomorphisms for a while and consider some utilitarian aspects of modular arithmetic. We are all familiar with its use in connection with time:

'Five hours later than 22.00 is 03.00' (mod 24)
'Today is Sunday: ten days from now will be Wednesday' (mod 7)
'Christmas Day is on a Tuesday this year so it will be on a
   Wednesday next year' (365=1 mod 7).

Distance-meters in cars also come to mind:

'The journey ahead is about 320 miles. The last three figures are now 834, so when they read 154 I should be nearly home'.

(On a long journey the driver often works on a three figure basis, ignoring the rest, thus using 1,000 as modulus.)
   The processes of addition and multiplication to a specified modulus can be represented on clocks:

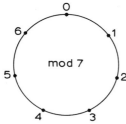

Fig. 1.1

$$5+3=1 \text{ (mod 7)},$$
$$6+4=3 \text{ (mod 7)},$$
$$4\times5 \text{ (mod 6)}=4 \text{ steps of } 5=2 \text{ (mod 6)},$$
$$3\times2 \text{ (mod 6)}=0 \text{ (mod 6)}.$$

1.8 Subtraction in modular arithmetic

So far we have only discussed the processes of addition and multiplication in relation to moduli. Subtraction is defined as the inverse of addition and presents no special difficulty.

For example, with modulus 5

$$4+2=1.$$

It follows that

$$1-4=2$$

and

$$1-2=4.$$

9

Verifying these statements for particular cases:

$$21\,(1 \bmod 5) -14\,(4 \bmod 5) = 7\,(2 \bmod 5),$$
$$6\,(1 \bmod 5) -12\,(2 \bmod 5) = -6\,(4 \bmod 5).$$

The statement $1-4=2 \pmod 5$ can be expanded into the statement: 'For modulus 5, any member of class 4 subtracted from any member of class 1 yields a member of class 2'. Results can conveniently be read from a table:

+	0	1	②
0	0	1	2
1	1	2	0 ↙
2	2	0	1

With modulus 3, the arrow points to the appropriate place for reading the value (0–1) to give a result 2 (encircled). It may be noted here that in this arithmetic $-2, -5, -8, \ldots$ all belong to class 1. In later paragraphs we shall be thinking of the set of integers rather than the set of natural numbers. If it suits us to replace 3 by $-4 \pmod 7$ we shall do so. Either number may be chosen as representative of its class.

1.9 Division in modular arithmetic

Consider the table for multiplication modulo 5:

×	0	1	2	3	4
0	0	0	0	0	0
1	0	1	2	3	4
2	0	2	4	1	3
3	0	3	1	4	2
4	0	4	3	2	1

From the table $\qquad\qquad 4 \times 3 = 2.$
The inverse statement is $\qquad 2/4 = 3.$
Does this statement expand into 'any number in class 2 divided by any number in class 4 yields a number in class 3'? The answer is 'yes and no'.

'No' because, for example, 17/9 is not equal to 3 or any integer in class 3
'Yes' because we can certainly find a solution to the congruency.

$$9x \equiv 17 (\text{mod } 5). \qquad (1.2)$$

First it reduces to

$$4x \equiv 2 \,(\text{mod } 5)$$

and then if we multiply both sides by the reciprocal of 4, i.e. 4 (see paragraph 1.12 for a definition of reciprocal),

$$16x \equiv 8 \,(\text{mod } 5),$$
$$x \equiv 3 \,(\text{mod } 5).$$

This solution can be verified by substitution in equation (1.2):

$$9 \times 3 \equiv 27, \text{ which is congruent to } 17.$$

1.10 Finding solutions to a congruence

We have already seen in section 1.9 how to solve one linear equation (more precisely a linear congruence). How many solutions are there to the equations $x^2 = 4$; $x^2 = 3$, among the five elements 0, 1, 2, 3, 4 in arithmetic modulo 5?

The table supplies the answers to both these equations: for the first $x = 2$ or 3; for the second there is no solution. The cubic equation $x^3 = 3$ has only one solution $x = 2$, but the equation $x^3 = x$ has 3 solutions 0, 1, and 4. All this is remarkably like ordinary algebra when we look for real roots to equations. The possible number of solutions seems to depend on the degree of the equation. However the reader should be warned that this conformity with ordinary algebra only exists when a prime modulus is used. For example the equation $x^2 = 1$, though of the second degree, has four different solutions modulo 8, namely 1, 3, 5, and 7.

To take another example here are three different ways of solving the equation

$$x^2 + 5x - 1 = 0 \,(\text{mod } 7),$$

all leading to the same solutions 4 and 5.

$$\begin{aligned}
\text{(a)} \quad & x^2 + 5x - 1 = 0 \,(\text{mod } 7) \\
& x^2 + 5x + 6 = 0 \\
& (x+3)(x+2) = 0 \\
& x = -3 \text{ or } -2 \\
& x = 4 \text{ or } 5.
\end{aligned}$$

11

(b) $x^2 + 5x - 1 = 0 \pmod{7}$
$$x^2 - 2x = 1$$
$$x^2 - 2x + 1 = 2$$
$$(x-1)^2 = 9$$
$$x - 1 = 3 \text{ or } -3$$
$$x = 4 \text{ or } 5.$$

(c) Finally applying the well-known formula
$$x = \{ -b \pm \sqrt{(b^2 - 4ac)} \}/2a$$
$$x = \{ -5 \pm \sqrt{(25 + 4)} \}/2$$
$$x = (-5 \pm \sqrt{1})/2$$
$$= -4/2 \text{ or } -6/2$$
$$= 3/2 \text{ or } 1/2.$$

Referring to Table 1.1(b) below, $3/2$ may be interpreted (row 2 under element 5) as 5 and $1/2$ as 4. It is one of the charms of working to a prime modulus that every expression which looks like a rational number (of the form p/q where p and q are integers and $q \neq 0$) may be replaced by an integer.

Exercises

1.8   Solve the equations
$$x^2 = 2, \ x^3 = 2, \ x^3 = 6$$
where x stands for one of the residue classes (mod 7).

1.9   Find all possible values for x in each of the following cases:

(a) $x^2 + 4x = 0 \pmod{11}$
(b) $x^2 = 4 \pmod{12}$
(c) $x^3 + 5x^2 - 6x = 0 \pmod{11}$
(d) $x^2 + 1 = 0 \pmod{2}$.

1.11   The identity element

Question 1.6 in section 1.6 is a crucial one. If you have not yet discovered the answer here it is. Each of the two sets of elements contains a very special one. For addition, mod 4 the special element is 0: it has the property of making no change when combined with any other element, e.g. $0 + 3 = 3$ and $2 + 0 = 2$. The special element for multiplication is 1 and this too has the 'no change' property. Such an element is commonly called an *identity element* or a *neutral element*. To display the isomorphism between the two sets of four elements it is necessary to match the identity elements with each other in setting out the correspondences.

12

1.12 Additive inverse and reciprocal

If $x+y=0$, then y is said to be the *additive inverse* of x.
If $xy=1$, then y is said to be the *reciprocal* of x.

Both the additive inverse and the reciprocal, where it exists, may readily be obtained from the appropriate tables. It is important to notice, however, that if a non-prime modulus is used some numbers may have no reciprocal. For modulus 6, 1 is its own reciprocal and so is 5 but the other elements have no reciprocal. There is, as we would expect, no reciprocal of zero with any modulus.

1.13 Arithmetic modulo 7 – Isomorphism

We are now in a position to tackle question 1 of section 1.6. We shall look for an isomorphism between multiplication modulo 7 and addition modulo 6. First compare the tables:

Table 1.1(a) (Mod 6)

+	0	1	2	3	4	5
0	0	1	2	3	4	5
1	1	2	3	4	5	0
2	2	3	4	5	0	1
3	3	4	5	0	1	2
4	4	5	0	1	2	3
5	5	0	1	2	3	4

Table 1.1(b) (Mod 7)

×	1	2	3	4	5	6
1	1	2	3	4	5	6
2	2	4	6	1	3	5
3	3	6	2	5	1	4
4	4	1	5	2	6	3
5	5	3	1	6	4	2
6	6	5	4	3	2	1

Table (b) does not look like table (a) but we have hopes of rearranging the elements to produce the same diagonal pattern. Let us, if possible, build up a table gradually by putting 1 into correspondence with 0 and choosing another element 2 to follow it. Successive steps are shown below:

×	1	2
1	1	2
2	2	4

×	1	2	4
1	1	2	4
2	2	4	
4	4		

×	1	2	4
1	1	2	4
2	2	4	1
4	4	1	2

13

This clearly will not do: the element 1 has recurred much too soon. Since we are using a trial and error method we will discard 2 as our second element and choose 3 instead. Successive steps give the following:

×	1	3
1	1	3
3	3	2

×	1	3	2
1	1	3	2
3	3	2	6
2	2	6	

×	1	3	2	6
1	1	3	2	6
3	3	2	6	4
2	2	6	4	
6	6	4		

leading to a final result:

×	1	3	2	6	4	5
1	1	3	2	6	4	5
3	3	2	6	4	5	1
2	2	6	4	5	1	3
6	6	4	5	1	3	2
4	4	5	1	3	2	6
5	5	1	3	2	6	4

The characteristic cyclic pattern is present once again and the correspondence with the elements for addition modulo 6 can be set out:

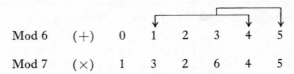

Mod 6   (+)   0   1   2   3   4   5

Mod 7   (×)   1   3   2   6   4   5

Checking particular cases for isomorphism:

(a)   From columns 2 and 5

$1+4=5 \pmod 6$
$3 \times 4 = 5 \pmod 7$.

14

This is a consistent result since 5 is a self-corresponding element in the two systems.

$$\text{(b)} \quad 6^3 \, (\text{mod } 7) = 216 \, (\text{mod } 7)$$
$$= \quad 6 \, (\text{mod } 7).$$

The corresponding calculation in the additive system is

$$3 \times 3 \, (\text{mod } 6) = 3 \, (\text{mod } 6)$$

and the two results 6 (mod 7) and 3 (mod 6) also correspond.

1.14 Automorphism

We set out to look for an arrangement of the six elements 1, 2, 3, 4, 5, 6 in multiplication modulo 7 which would be isomorphic to the six elements of addition modulo 6 in the order 0, 1, 2, 3, 4, 5. We had to reject one arrangement beginning 1, 2, . . . but were successful with an arrangement beginning 1, 3, . . . Could there be any other successful arrangement beginning 1, 4, . . . or 1, 5, . . . or 1, 6, . . .? Further application of the trial and error method would in fact reveal one other successful arrangement 1, 5, 4, 6, 2, 3. We now have what might be called a triple isomorphism.

Addition	(mod 6)	0	1	2	3	4	5
Multiplication	(mod 7)	1	3	2	6	4	5
Multiplication	(mod 7)	1	5	4	6	2	3

Testing a particular case once more, in the 4th and 5th column $3+4=1$ (mod 6), $6 \times 4 = 3$ (mod 7), $6 \times 2 = 5$ (mod 7). This result is confirmed in the second column which shows a correspondence between 1 mod 6, 3 mod 7, and 5 mod 7 in the three systems of 6 elements.

Considering the 2nd and 3rd rows alone, we have found an example of *automorphism* within the set of six elements for multiplication (mod 7). The six elements can be put into isomorphic correspondence with another arrangement of themselves. We shall look at other examples of automorphisms amongst residues to a prime modulus in Chapter 3.

15

1.15 Multiplication Modulo 6

It is now time to discuss the multiplication table for composite numbers such as 4, 6, 10 (see exercise 1.7).

Here is the table for modulus 6:

×	0	1	2	3	4	5
0	0	0	0	0	0	0
1	0	1	2	3	4	5
2	0	2	4	0	2	4
3	0	3	0	3	0	3
4	0	4	2	0	4	2
5	0	5	4	3	2	1

The border of zeros is familiar but the 5×5 array is strikingly different from those which appeared for the prime moduli 3, 5, 7.

(i) It is not true that each element appears once in every row and column.

(ii) Zeros appear in three of the rows: for example, $2 \times 3 = 0$.

Now if x and y are defined, say, as rational numbers, and it is known that $xy = 0$, it can be said at once that either x or y or both must be zero. This principle is used in the ordinary method of solving the quadratic equation $(x - 3)(2x - 1) = 0$, leading to $x = 3$ or $\frac{1}{2}$. But we have observed that if x and y are defined as residue classes modulo 6 and $xy = 0$, then *both x and y* may take non-zero values such as 2, 3 or 3, 4. The same could evidently be said in general of the residue classes modulo n where n is a composite integer. Such a system is said to have *factors of zero* or *zero divisors*. It does not qualify as a *field*, a term which we shall not define formally at present. It is an example of a system called a *ring* and we shall call this particular ring Z_6 following a commonly used notation. Other instances of rings will be discussed later in Chapters 3 and 7.

Exercises

1.11 Write out the table for multiplication mod 10. Which rows contain zeros? Which do not contain zeros?

1.12 Can you find a working rule for the appearance or otherwise of zeros?

16

1.13 Can you, without writing out the full table, forecast a row which will contain zeros for mod 12, and a row which will not contain zeros? Verify by writing out these rows in full.

We have now incidentally answered question 2 of section 1.6. 'Does it work for all moduli?' There is no isomorphism between the five non-zero residues in the multiplication system mod 6 and the five residues of the addition system mod 5. Such isomorphisms only exist when multiplication is considered in respect of a prime modulus.

1.16 Structures with four elements

When dealing with addition modulo 4 and multiplication modulo 5 we came across a cyclic system with 4 elements. To be more precise we made acquaintance with two representations of a cyclic *group* of order 4. No formal definition of a group will be given at this stage but we will summarise some of the features we noticed on the way:

(i) The group consisted of a set of elements and a way of combining two elements (addition in one case, multiplication in the other) to produce an element of the group.

(ii) Each element of the group occurred once in every row and column of the group table.

(iii) One element of the group (0 for addition and 1 for multiplication) had the special property of combining with every other element to produce the same element again:

$$0+2=2, 3+0=3; 1\times4=4, 2\times1=2.$$

Another feature that we noticed was that the group for multiplication at first looked somewhat different from the group for addition but on rearrangement of the elements was observed to have the same cyclic structure. The question arises whether all groups with four elements have exactly the same type of structure. To answer this question we shall turn to a geometrical example.

1.17 Symmetry

Here is a sketch of an educational toy of the kind often given to a child four or five years old to help him develop his sense of space. A set of flat objects whose shapes are, say, a regular hexagon, a square, a

17

rectangle, a 'tree' and a 'cat', goes with a flat tray containing holes into which the various objects fit exactly.

Fig. 1.2

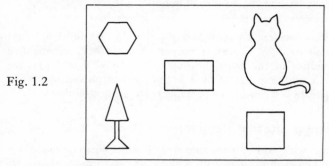

A child is likely to find from experience, though he would not clearly understand the reason, that some shapes are much easier to fit into the proper space than others. His father and mother would realise that this is because there are more ways of fitting the hexagon and square correctly than there are of fitting the tree or the cat. If the tray is green and the shapes are coloured green on one side only, then the number of ways of fitting the spaces to obtain an all-green surface may be listed thus:

cat 1, tree 1, rectangle 2, square 4, hexagon 6.

On the other hand, if both sides are green, the list becomes:

cat 1, tree 2, rectangle 4, square 8, hexagon 12.

The ease with which the shapes may be fitted clearly depends on the degree of symmetry possessed by each shape. The square for instance has *rotational* symmetry of order four which means that it can be given a quarter turn and again fit the space and also that if the quarter turn is repeated four times in succession it will return to the original position. In addition it has *bilateral* symmetry about four different axes named 1–4 in the diagram

Fig. 1.3

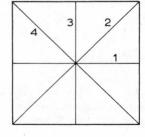

and this means that it can be given a half-turn about any of these axes and still fill the space.

By way of comparison and contrast a swastika has the first kind of symmetry but not the second.

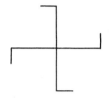

Fig. 1.4

The cat, which as drawn possesses no symmetry, can be correctly placed in one position only and will not fit the hole in any position if turned over.

Let us now consider in some detail the case of the rectangle, naming it ABCD and its axes of bilateral symmetry PQ and RS.

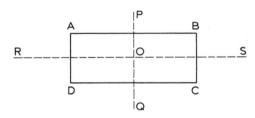

Fig. 1.5

We can by rotation about various axes transform the rectangle *so that it still occupies the same rectangular space but its vertices are in different positions.* We can rotate it 180° about PQ or RS, or about the line through O perpendicular to the plane of the rectangle. We could also leave it where it is. This makes four different ways of treating the rectangle:

(1) Leave it alone.
(2) Rotate it 180° about PQ.
(3) Rotate it 180° about RS.
(4) Rotate it 180° about a line through O
perpendicular to the plane ABCD.

19

The four possible orientations which result from applying these *transformations* in turn to the original rectangle are given below. Although drawn separately for convenience they should be visualised as occupying successively the same rectangular space.

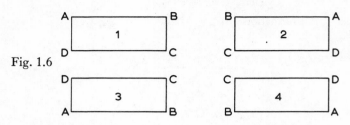

Fig. 1.6

Now each of the transformations (1), (2), (3), (4) can be applied to each of the rectangles 1, 2, 3, 4 and a table of results made out:

Transformations	*on rectangles*				
	1	2	3	4	
(1)	1	2	3	4	⎫
(2)	2	1	4	3	⎬ resulting
(3)	3	4	1	2	⎬ rectangles
(4)	4	3	2	1	⎭

This table should be read as follows:
Transformation (2) acting on rectangle 4 produces rectangle 3.
The table can, however, with advantage be modified into one showing the products of two transformations in *succession*. We have only to insert brackets as shown below:

Table 1.2
first transformation

S	(1)	(2)	(3)	(4)	
(1)	(1)	(2)	(3)	(4)	⎫
second (2)	(2)	(1)	(4)	(3)	⎬ equivalent
transformation (3)	(3)	(4)	(1)	(2)	⎬ single transformation
(4)	(4)	(3)	(2)	(1)	⎭

Now the table is read somewhat differently:
Transformation (2) succeeding transformation (4) produces the same result as transformation (3) when applied to the original rectangle (or indeed to any of the rectangles).

20

The table shows the properties of a *group of transformations* of order 4. We see again that:

(i) Each element of the group appears once in every row and column.

(ii) There is an identity element (1).

(iii) If two transformations (or the same transformation twice) are combined by succession they give an element of the group.

It remains to be investigated whether it has the same structure as the cyclic group of four elements for multiplication modulo 5 (table 1.3).

Table 1.3

×	1	2	4	3
1	1	2	4	3
2	2	4	3	1
4	4	3	1	2
3	3	1	2	4

Table 1.2 does not look like table 1.3 and the reader may verify that no rearrangement will make it so. If any two lines in 1.2 are compared each is seen to be derived from the other by interchanging the elements in pairs, and the same may be said for any two columns. This is quite unlike the cyclic shift observed in table 1.3. This new group of table 1.2 is named Klein's 4-group after the famous mathematician who is associated with it.

It is a well-known fact that there are only two groups of order 4 (possessing four elements) and they are the two considered in this chapter – the cyclic group and Klein's group.

Exercises

1.14 (a) Write out the multiplication table for residue-classes modulo 8. Cross out all the rows and columns containing zeros and make a new table using four elements only – the ones not crossed out.

Compare your result with tables 1.2 and 1.3. Does it resemble either? If so set out the isomorphism in detail and check it in one or two instances.

(b) Explore the residue-classes mod 10 and mod 12 in the same way.

(c) Write out multiplication tables for the following sets of residue-classes using modulus 30 in each case.

$$\{1, 7, 19, 13\}, \{2, 4, 8, 16\} \{5, 25\}$$

In the first set 1 is an identity element leaving every element in the set unchanged when used as a multiplier. Have the last two sets got identity elements? Are the first two sets isomorphic for multiplication?

1.15 Two couples at a dinner-dance agree to rearrange the seating at their table after each dance according to one or other of four instructions to be drawn from a pool:

(a) Men change places with ladies on their right,
(b) Men change places with ladies on the left,
(c) Men change places, ladies change places,
(d) Return to the same seats as before.

This was the initial arrangement.

Fig. 1.7

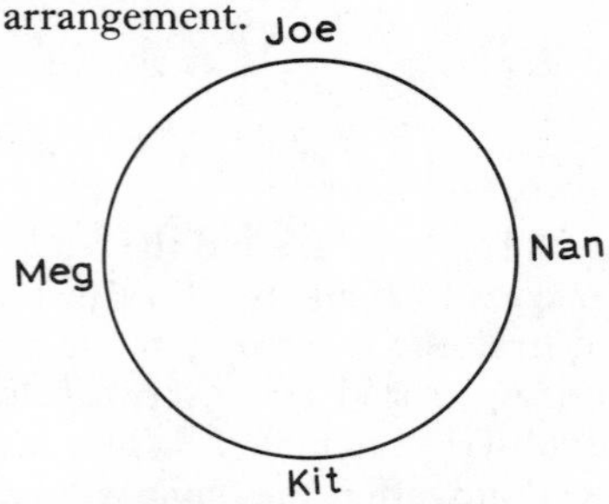

Do you think that the four instructions form a group? If so is it a Klein's group or cyclic?

1.18 Summary

So far, although we have used the term 'group' we have not formally defined it. We have explored combinations of numbers by the binary operations of addition and multiplication using various moduli, and also one example of transformations of a geometrical figure by means of rotations, so that it still occupies the same geometrical space. We found certain characteristics in common and the appearance of regular structures obeying definite rules. In the next chapter we shall consider various representations of two groups of 6 elements in order to gain further insight into the nature of a group.

GROUPS OF ORDER SIX:
formal definition of a group

2.1 Rotations of a regular hexagon

What rotations can we perform on a regular hexagon subject to the condition that afterwards it is to occupy the same hexagonal space? First consider the case with an additional constraint that the hexagon is not to move out of its plane. It is not difficult to see that there are six operations and six only which fulfil these two conditions:

(1) Leave it alone.
(2) Rotate anti-clockwise through 60°.
(3) Rotate anti-clockwise through 120°.
(4) Rotate anti-clockwise through 180°.
(5) Rotate anti-clockwise through 240°.
(6) Rotate anti-clockwise through 300°.

Clockwise rotations have been ignored because −60°, for instance, gives the same result as +300°. There are only six positions for the hexagon to occupy and rotations of −60°, +300°, +660° . . . all lead to the same one. Any one of them is typical of its class of rotation. Rotation (2) succeeding rotation (3) gives the same result as rotation (4). In symbols (2)S(3)=(4). A table giving the result of two successive rotations can be built up very rapidly:

Table 2.1

S	(1)	(2)	(3)	(4)	(5)	(6)
(1)	(1)	(2)	(3)	(4)	(5)	(6)
(2)	(2)	(3)	(4)	(5)	(6)	(1)
(3)	(3)	(4)	(5)	(6)	(1)	(2)
(4)	(4)	(5)	(6)	(1)	(2)	(3)
(5)	(5)	(6)	(1)	(2)	(3)	(4)
(6)	(6)	(1)	(2)	(3)	(4)	(5)

This pattern is exactly the same as for addition modulo 6. It may be recalled that modular addition was represented on a clock in section 1.7 and this in itself suggests rotations as an analogy. The two systems are isomorphic. Note that the rotation which produces no change has been called (1) so the table as written suggests multiplication rather than addition for which the identity element is 0. In connection with groups the word multiplication is often used in a very broad sense to cover a variety of situations. Where there is no danger of ambiguity a $*b$ is often written ab even though the symbol $*$ might stand for addition or succession.

Second, if rotations out of the plane were permitted then the hexagon could be reversed by giving it a half-turn about any one of its 6 axes of symmetry. A group of 12 rotations would emerge; but rather than study so large a group at this stage it is better to look at a similar but smaller group with only 6 elements.

2.2 The equilateral triangle: a group of transformations

An equilateral triangle ABC can be transformed in just six ways so that it occupies the same triangular space though its vertices occupy different positions.

(a) Rotations of $0°$, $120°$, $240°$ about an axis through its centre O, perpendicular to its plane.

(b) Reflections in its axes of symmetry OA, OB, OC.

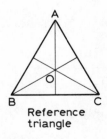

Reference
triangle

Fig. 2.1

The set of six possible orientations of the triangle have been drawn below and numbered. As arranged they look like two sets of 3 triangles which are images of each other in a mirror m. In actual fact each triangle would occupy the same triangular space as the reference triangle.

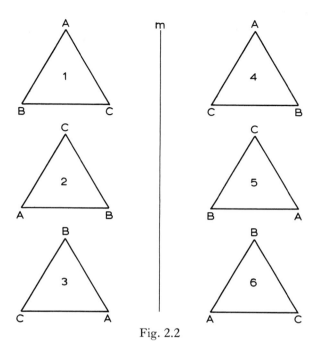

Fig. 2.2

Corresponding transformations which would bring the reference triangle to these positions are:

       (1) No change.
       (2) Rotate 120° anti-clockwise.
       (3) Rotate 240° anti-clockwise.
       (4) Reflect in OA.
       (5) Reflect in OB.
       (6) Reflect in OC.

To understand the structure of the group it is useful to consider the effect of each transformation on the complete set of orientations. 'Reflect in OA' is to be understood as reflection in a fixed mirror along the line OA of the reference triangle.

(1) Produces no change.

(2) Produces a cyclic effect 1——→2, 2——→3, 3——→1 on the first subset and a reverse effect 4——→6——→5 on the second subset which is 'through the looking glass'.

(3) is similar in effect to (2) producing 1——→3——→2 and 4——→5——→6.

(4), (5), and (6). The reflective transformations produce interchanges as follows:

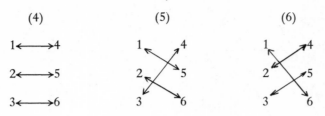

(4)   (5)   (6)

All of this can be summed up in a table of results:

| | triangles | | | | | |
	S	1	2	3	4	5	6
	(1)	1	2	3	4	5	6
	(2)	2	3	1	6	4	5
transformations	(3)	3	1	2	5	6	4
	(4)	4	5	6	1	②	3
	(5)	5	6	4	3	1	2
	(6)	6	4	5	2	3	1

resulting triangles

This is not yet a group-table but is a very useful stepping-stone on the way. It is only necessary to replace 1 by (1), 2 by (2), etc., to convert it into the 'multiplication table' of a group of transformations similar to table 2.1. As it stands, for example, the table tells us that transformation (4) acting on triangle 5 produces triangle 2: but also after modification the new table has a different interpretation, equally valid, that transformation (4) succeeding transformation (5) has the same final effect on any of the triangles as transformation (2). In symbols

$$(4)\,S\,(5)=(2).$$

Although this statement is true for all triangles the reader will find in practice that it is more convenient to use triangle 1 (the reference triangle) to verify any entry in the revised table below:

Table 2.2
first transformation

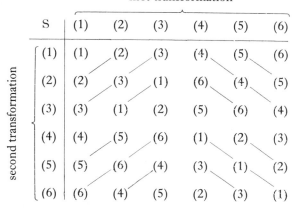

S	(1)	(2)	(3)	(4)	(5)	(6)
(1)	(1)	(2)	(3)	(4)	(5)	(6)
(2)	(2)	(3)	(1)	(6)	(4)	(5)
(3)	(3)	(1)	(2)	(5)	(6)	(4)
(4)	(4)	(5)	(6)	(1)	(2)	(3)
(5)	(5)	(6)	(4)	(3)	(1)	(2)
(6)	(6)	(4)	(5)	(2)	(3)	(1)

(second transformation labels the rows)

2.3 Definition of a Group

The axioms which define a group will now be stated using the examples already discussed for illustration.

A set of elements a, b, c . . . together with a rule for combining any two of them, symbolised by $*$ forms a group G if the following four conditions are satisfied:

Axiom I *Closure:* for all a, b in G, $a * b$ is a member of G.

Axiom II *Associative Property:* for all a, b, c in G $(a * b) * c = a * (b * c)$.

Axiom III *Identity Property:* there exists a special element e in G such that, for all a in G, $a * e = a$ and $e * a = a$.

Axiom IV *Inverse Property:* for each a in G there exists a unique element a^{-1} in G such that $a * a^{-1} = e$ and $a^{-1} * a = e$.

The general name for a^{-1} is 'inverse of a' but when the rule of combination is ordinary multiplication it is often referred to as the 'reciprocal of a' while for addition it may be called 'the negative of a'.

2.4 Illustrations of the axioms

(a) Residues modulo 5

This system was discussed informally in section 1.3. The non-zero elements 1, 2, 3, 4 form a group of order 4 when the rule of combination is multiplication. The group table shows that closure is obeyed. The identity element is 1 and the inverses of 1, 2, 3, 4 respectively are 1, 3, 2, 4 because, for example, $4 \times 4 = 1$. Finally the system possesses the associative property. This is because the property is widely held by many different systems. Amongst these is the set of integers which possesses the property both for addition and multiplication. Hence the residue classes derived from the integers also possess the property for both rules of combination.

The name of the group is the cyclic group of order 4, and the usual symbol for it is C_4; other examples of it are residue classes, mod 4 (addition) and rotations of a square in its own plane (succession).

(b) Transformations of the equilateral triangle

The six transformations (1) . . . (6) discussed in section 2.2 form a group of order 6 when combined by succession. (By contrast the six orientations of the triangle do not form a group because there is no way of combining two triangles to make a third.) The identity element is 'leave the triangle unchanged'. The inverse of (3) is (2) while (4) is self-inverse. This is an example of a non-commutative group since (4) S (5)=(2) and (5) S (4)=(3). It is the reflective nature of the transformations (4), (5), and (6) which makes the group non-commutative by reversing the direction of the cyclic shift in the right-hand half of the group table. The name of this group is the dihedral group of order 6 and the usual symbol for it is D_3. It is easy to see that exactly the same group would arise if we substituted half-turns about OA, OB, OC for the reflections. This would bring the other face of the triangle uppermost and hence the name dihedral which means two-faced. A square rotated through $0°$, $90°$, $180°$, and $270°$ and also given half-turns about its four axes of bilateral symmetry would give rise to D_4, the dihedral group of order 8.

Exercises

Refer to table 1.2 in section 1.17.

2.1 Find the inverse of each element in the group.

2.2 Is the group commutative?

2.3 Verify that $\{(2)S(3)\} S (4) = (2)S\{(3)S(4)\}$
and that $\{(2)^2\}S(4) = (2)S\{(2)S(4)\}$ where $(2)^2$ is understood as $(2)S(2)$.

2.5 Use of the axioms to solve an equation relating elements in a group

Let $ax=b$ where a, b, and x are all members of a group whose identity element is e. It is often convenient to use an alternative form of this relation which isolates x. The alternative form is $x=a^{-1}b$ and it is derived from the first relation by use of the group axioms as follows:

$$ax= b,$$
$a^{-1}(ax)=a^{-1}b$, every element a has an inverse a^{-1},
$(a^{-1}a)\,x=a^{-1}b$, associative property,
$e\,x=a^{-1}b$, by definition of inverse,
$x=a^{-1}b$, property of the identity element.

Exercises

2.4 Here is the table for the non-commutative group D_3:

	p	q	r	u	v	w
p	p	q	r	u	v	w
q	q	r	p	w	u	v
r	r	p	q	v	w	u
u	u	v	w	p	q	r
v	v	w	u	r	p	q
w	w	u	v	q	r	p

Verify the associative property in the following cases:

(a) $(uq)\,w=u\,(qw)$
(b) $(q^2)v=q\,(qv)$
(c) any other instance of your own choice.

2.5 Referring to the same table:

(a) Which is the identity element?
(b) Which element is the inverse of q?
(c) Replace $q^{-1}v\,q$ by a single element and find also the single element which is the equivalent of $q\,u\,q^{-1}$.

2.6 Use the method of section 2.5 to show that in any group whatever if $m^{-1}k\,m=j$ then $m\,j\,m^{-1}=k$.

2.6 The cancellation property

In ordinary algebra if $ax=bx$ and $x \neq 0$, it follows that $a=b$, and it is common practice to speak of cancelling x in these circumstances. There is a similar property in group algebra which can quickly be derived from the group axioms of section 2.3 and may even be used as an alternative to both of axioms III and IV. The property may be stated thus:

If a, b, x are elements of a group G and if either $ax=bx$ or $xa=xb$ then $a=b$. (Only if G is known to be a commutative group would it be permissible to cancel x when $ax=xb$.)

Here is the proof for the first case:

$$ax=bx,$$
$$(ax)x^{-1}=(bx)x^{-1} \text{ (inverse axiom)},$$
$$a(xx^{-1})=b(xx^{-1}) \text{ (associative axiom)},$$
$$ae=be \text{ (inverse axiom)},$$
$$a=b \text{ (identity axiom)}.$$

If it were possible for ax to be equal to bx and $a \neq b$ this would mean that under heading x of the group table there would be some element which occurred twice in the column. Likewise if $xa=xb$ and $a \neq b$ some line in the table would have a repeated element. Hence the cancellation property confirms what we have already observed that for a finite group each element occurs once and only once in every line and column of the multiplication table.

2.7 The associative property extended

The associative property of groups prevents there being any ambiguity in the product abc of three elements a, b, and c. There is a well-established proof by induction that the property can be extended to four or more elements. Given a product of n elements it is permissible to partition it by brackets wherever we please provided we do not tamper with the sequence of the elements.

Thus ab^2cde may be interpreted as $\{(ab)(bc)\}$ (de) or as $\{a(b^2)\}$ $\{c(de)\}$.

We shall need this property in later chapters.

Exercise

2.7   In table 2.4 find qr, ru, uv and verify that $(qr)(uv)=q\{(ru)v\}=\{q(ru)\}v$. Also verify that $\{(ur)^2\}(v^3)=\{u(ru)\}\{(rv)v^2\}$.

2.8 The inverse of a product

It is often necessary to find the inverse of the product of two or more elements and the extended associative property is immediately useful for this purpose. Suppose we need the inverse x of ab; then by the inverse property

$$(ab)x = e,$$
$$(b^{-1}a^{-1})\{(ab)x\} = (b^{-1}a^{-1})e,$$
$$b^{-1}(a^{-1}a)bx = b^{-1}a^{-1},$$
$$b^{-1}ebx = b^{-1}a^{-1},$$
$$x = b^{-1}a^{-1}.$$

It is left to the reader to show by a similar method that

$$(abc)^{-1} = c^{-1}b^{-1}a^{-1}.$$

2.9 Infinite groups

Although this book is mainly concerned with finite structures it is important to state at this stage that infinite sets of elements may also comply with the group axioms for an appropriate rule of combination. Good examples of infinite groups are:

(1) the integers under addition;

(2) terms of the sequence $\dots, \frac{1}{4}, \frac{1}{2}, 1, 2, 4, 8, \dots$ for multiplication;

(3) rational numbers (of the form p/q where p and q are integers and $q \neq 0$) for addition;

(4) rational numbers, excluding zero (p also $\neq 0$), for multiplication.

It does not take long to verify that they conform to the four axioms of a group. Since all numbers have the associative property for addition and multiplication this may be taken for granted in each case. Note that the integers do not form a group for multiplication because, for example, the integer 3 has no inverse. Its reciprocal $1/3$ might seem to qualify at first sight but then $1/3$ is not an integer so it is excluded from consideration.

We have already seen instances in which an infinite group can be treated as if it were finite by partitioning it into classes, e.g. the integers may be partitioned into even and odd integers. The simple scheme in section 1.1 can then be regarded as a summary of operations performed over the whole set of integers. In multiplying even by even we regard ourselves as performing a vast multiplication of every even number by every even number. We shall meet again this idea of extended multiplication applied to classes of elements within a group.

31

2.10 Isomorphism

A group having been formally defined it is advisable also to say here precisely what is meant by the expression 'isomorphism between two groups'. A group G with elements $a, b, c, \ldots$ and a rule of combination $*$ is said to be isomorphic with another group G' whose elements are $a', b', c', \ldots$ and rule $\otimes$ if the elements of the two systems can be put into one-one correspondence with each other: $a \longleftrightarrow a'$, $b \longleftrightarrow b'$, $\ldots$ in such a way that for all possible choices of a and b

$$(a * b) \longleftrightarrow (a' \otimes b').$$

The elements a and a' are said to be images of each other. Isomorphism is a mutual relation.

It is often convenient to resolve the one-one correspondence mentioned in the definition and symbolised by a double arrow into two one-way *mappings* symbolised by a single arrow:

$$a \longrightarrow a',$$
$$b \longrightarrow b',$$

and
$$(a * b) \longrightarrow (a * b)'.$$

With this notation a' is said to be the image of a under the mapping and one aspect of the isomorphism is then expressed symbolically as

$$(a * b) \longrightarrow (a' \otimes b'),$$

or otherwise as
$$a' \otimes b' = (a * b)'. \tag{2.1}$$

In the latter form it may be expressed in words as 'the product of the images is the image of the product'. Here product refers impartially to the result of using either of the two rules of combination.

Another way of expressing the same property is to give the symbol φ to the mapping and the symbol $\varphi(a)$ to the image of a. Equation (2.1) in this notation becomes

$$\varphi(a) \otimes \varphi(b) = \varphi(a * b).$$

Diagrammatically the situation looks like this:

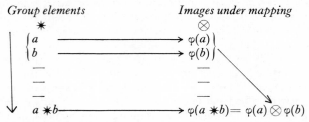

The key property of the isomorphism is that the same element in the right hand column is arrived at by two different paths: (a) as image of the product $a * b$, and (b) as product of the images $\varphi(a)$ and $\varphi(b)$.

To take a particular example, we have already considered in section 1.4 the isomorphism which exists between residue classes 0, 1, 2, 3 for addition mod 4, and the residue classes 1, 2, 4, 3 for multiplication mod 5. The one-way mapping φ is:

$$
\begin{array}{ccc}
\text{Mod } 4(+) & & \text{Mod } 5(\times) \\
0 & \longrightarrow & 1 \\
1 & \longrightarrow & 2 \\
2 & \longrightarrow & 4 \\
3 & \longrightarrow & 3
\end{array}
$$

Under this mapping $\varphi(0)=1$, $\varphi(1)=2$, $\varphi(2)=4$, and $\varphi(3)=3$.

Since the correspondence in an isomorphism is essentially two-way there is an inverse mapping φ^{-1}, namely:

$$
\begin{array}{ccc}
\text{Mod } 5(\times) & & \text{Mod } 4(+) \\
1 & \longrightarrow & 0 \\
2 & \longrightarrow & 1 \\
4 & \longrightarrow & 2 \\
3 & \longrightarrow & 3
\end{array}
$$

In examples 1–5 of section 1.4 we used both mappings indiscriminately in order to pass from one system of elements to the other.

2.11 Automorphism

The term automorphism already used in section 1.14 will be formally defined in terms of mappings in chapter 3, but it is worth noting here that an automorphism exists in the multiplication system modulo 5 because an isomorphic mapping exists between the elements of the group and the same elements in a different order. The correspondences are as follows:

$$
\begin{array}{ccc}
1 & \longleftrightarrow & 1 \\
2 & \longleftrightarrow & 3 \\
4 & \longleftrightarrow & 4 \\
3 & \longleftrightarrow & 2
\end{array}
$$

2.12 Another representation of the Group D_3

It can easily be seen that the six possible permutations on three symbols are another illustration of the Group D_3. The analogy with the triangular transformations can be briefly indicated.

The six possible arrangements of the three symbols are:

1. *abc*,   4. *acb*,
2. *cab*,   5. *cba*,
3. *bca*,   6. *bac*.

There are six corresponding transformations or permutations:
(1) No change.
(2) Replace first symbol by third, third by second, second by first.
(3) Replace first symbol by second, second by third, third by first.
(4) Interchange second and third.
(5) Interchange first and third.
(6) Interchange first and second.

In table 2.2, $(2)S(5)=(4)$. We read this now as permutation (2) succeeding permutation (5) has the same effect as permutation (4).

Starting with the initial arrangement abc: (5) changes abc to cba, (2) changes cba to acb, also (4) changes abc to acb. Hence $(2)S(5)=(4)$.

It will be found that the whole of table 2.2 can be interpreted in this way as the 'multiplication table' for the system of permutation, thus showing that it is isomorphic with the groups of rotations of the equilateral triangle.

2.13 An algebraic illustration of D_3

A much less obvious representation of the same abstract group is the set of algebraic expressions below when combined by substitution.

$$\left.\begin{array}{ll} e_1=t, & e_4=1-t, \\ e_2=1-1/t, & e_5=1/t, \\ e_3=1/(1\text{-}t), & e_6=t/(t-1), \end{array}\right\} (2.2)$$

The multiplication table is obtained from combining two expressions by substitution, for which we will take the symbol $\circledS$.

For example ,
$$e_4 \circledS e_5 = 1-1/t = e_2,$$
$$e_2 \circledS e_3 = 1-1/\{1/(1-t)\}$$
$$= 1-(1-t)$$
$$= t$$
$$= e_1.$$

Any combination of two expressions in this way leads to one or other of the expressions and to the following table:

Table 2.3

$\circledS$	e_1	e_2	e_3	e_4	e_5	e_6
e_1	e_1	e_2	e_3	e_4	e_5	e_6
e_2	e_2	e_3	e_1	e_6	e_4	e_5
e_3	e_3	e_1	e_2	e_5	e_6	e_4
e_4	e_4	e_5	e_6	e_1	e_3	e_2
e_5	e_5	e_6	e_4	e_2	e_1	e_3
e_6	e_6	e_4	e_5	e_3	e_2	e_1

34

Comparison with table 2.2 shows at once that the system is isomorphic with the group of rotations of the equilateral triangle.

2.14 Another method of treatment

Another and possibly more illuminating way of looking at this representation of D_3 is to separate two different aspects of the algebraic forms. Each may be regarded as a passive expression or as an active transformation. As transformations they may be relabelled $t_1, t_2 \ldots t_6$ where, for example,

t_1 is the transformation $t \longrightarrow t$,
t_2 is the transformation $t \longrightarrow 1-(1/t)$,
t_5 is the transformation $t \longrightarrow 1/t$.

As expressions we can mark the difference by a change of symbol from t to a:

$$e_1 = a, e_2 = 1 - 1/a. \ldots$$

The effect of the transformation t_5 on these expressions may then be set out in detail as follows:

$$a \longrightarrow 1/a, \qquad\qquad 1-a \longrightarrow 1/(1-a),$$
$$1-1/a \longrightarrow a/(a-1), \qquad\qquad 1/a \longrightarrow a,$$
$$1/(1-a) \longrightarrow 1-a, \qquad\qquad a/(a-1) \longrightarrow 1-(1/a).$$

The transformations can be combined by succession S and the product obtained by studying the final effect on a. Thus $t_5 S t_2$ sends a first to $1-1/a$ and then sends $1-1/a$ to $a/(a-1)$ so that

$$t_5 S t_2 = t_6.$$

The isomorphism with the group of rotations of an equilateral triangle is now clearer. The transformations $t_1 - t_6$ correspond to the rotations, and the expressions to the orientations of the triangle.

The desirability of distinguishing between the active and passive elements in the system will again become evident in later work on substitution groups in chapter 10. We can make the distinction even clearer by replacing a in the expressions by $\cos^2\theta$. The set of expressions then becomes:

$$\cos^2\theta, \qquad\qquad \sin^2\theta,$$
$$-\tan^2\theta, \qquad\qquad \sec^2\theta,$$
$$\operatorname{cosec}^2\theta, \qquad\qquad -\cot^2\theta.$$

The passive objects of the transformations no longer have the same algebraic form as the transformations. It is of interest that all six of the trigonometric ratios emerge after the change of symbol and that they can be permuted among themselves by the six transformations. The effect of t_5 can be concisely expressed as:

$$\cos^2\theta \longleftrightarrow \sec^2\theta, -\tan^2\theta \longleftrightarrow -\cot^2\theta, \operatorname{cosec}^2\theta \longleftrightarrow \sin^2\theta.$$

Exercises

2.8 Here are twelve expressions which form a group for substitution. They are not arranged in an orderly sequence. $(2t-1)/(t+1)$, $(1-2t)/(2-t)$, t, $(t+1)/(2-t)$, $1/(1-t)$, $(2-t)/(t+1)$, $t/(t-1)$, $1/t$, $(t+1)/(2t-1)$, $1-t$, $1-1/t$, $(2-t)/(1-2t)$.

(a) Write out a set of transformations which may be derived from these expressions.

(b) Which is the identity transformation?

(c) Verify that the transformations when applied in turn to the number 3 produce the following set of numbers:

$$5/4, 5, 3, -4, -1/2, -1/4, 3/2, 1/3, 4/5, -2, 2/3, 1/5.$$

(d) Verify in at least four cases that any of the twelve transformations when applied to any of these numbers, produces no new number.

(e) Find the effect of the transformation $t \longrightarrow (1-t)$ on all the numbers mentioned in (3).

(f) Find the effect of repeated application of the transformation $t \longrightarrow (t+1)/(2-t)$ to the number 3.

(g) Use your results in (e) and (f) to put the transformations into an orderly array like that of the algebraic expressions e, to e_6 in the first part of this section. Number them $t_1, t_2 \ldots t_{12}$.

(h) Convert the twelve given expressions in t to expressions in a, e.g. change $(2t-1)/(t+1)$ to $(2a-1)/(a+1)$. Number the expressions a_1, $a_2, \ldots a_{12}$ so that they correspond with $t_1, t_2 \ldots t_{12}$. Examine the effect of applying each transformation in turn to $a_1, a_2, \ldots a_{12}$. Draw arrow patterns of results similar to those in section 2.2.

(j) Write out some typical rows of the multiplication table for the twelve transformations.

(k) The transformations are isomorphic with twelve rotations of a planar geometric figure. Which figure? What rotations? Write out a set of possible correspondences between transformations and rotations in detail.

2.9 (a) Explore the group of transformations below:

$$t \longrightarrow t, \qquad\qquad t \longrightarrow 2(t-1)/t,$$
$$t \longrightarrow (2-2t)/(2-t), \qquad t \longrightarrow 2-t,$$
$$t \longrightarrow (t-2)/(t-1), \qquad t \longrightarrow t/(t-1),$$
$$t \longrightarrow 2/t, \qquad\qquad t \longrightarrow 2/(2-t).$$

(b) Find a group of rotations which is isomorphic with this group of transformations.

(c) Show that the curve defined by the equations $x=t$, $y=(t-2)/(t-1)$, is the same as that defined by $x=2/(2-t)$, $y=2(t-1)/t$. Also explore the curve defined by $x=t$, $y=2(t-1)/t$.

Make a sketch of both curves. Are there other such curves?

2.15 Cross-ratio

The six expressions in t of section 2.12 appear in projective geometry. The cross-ratio of a linear range of four points A, B, C, D is defined as $\dfrac{AB}{BC} \Big/ \dfrac{AD}{DC}$ or in an alternative form as $(AB.CD)/(AD.CB)$. This number is written (ABCD) and has the remarkable property of being invariant when the range is projected. The points on the range can be named in any one of twenty-four different orders, another one for example being B, C, A, D. It can be proved that the twenty-four corresponding cross-ratios fall into six sets of four equal numbers, and that if one of these numbers is t, then the other five are $1-(1/t)$, $1/(1-t)$, ... It would be wearisome to prove this in detail but the reader may like to check that (BADC), (CDAB), and (DCBA) are all equal to (ABCD); and that if, for example, the transformation $(1-t)$ is applied to (ABCD) it becomes (DBCA).

2.16 Subgroups

It may have been noticed that table 2.3 has been divided into four compartments. This is to draw attention to an important feature of the group's structure: it contains subgroups. Disregarding e_4, e_5, e_6, the expressions e_1, e_2, e_3, form a cyclic subgroup of their own under the same rule of combination S as for the main group. A similar subgroup is apparent in table 2.2 as (1), (2), (3), where it represents the group of rotations of the triangle in its own plane.

It would be tempting to say that the set of expressions e_4, e_5, e_6 also forms a subgroup, but this would be a misnomer because there is no identity element in this set and closure is not obeyed. We therefore speak of them collectively as a coset of the subgroup $\{e_1, e_2, e_3\}$.

There are three other subgroups, all of order 2. A typical one is $\{e_1, e_4\}$:

S	e_1	e_4
e_1	e_1	e_4
e_4	e_4	e_1

37

Does the cyclic group of six elements also contain subgroups? Let us rewrite 2.1 in a different order, showing $\{(1), (3), (5)\}$ as a subgroup:

Table 2.4

s	(1)	(3)	(5)	(4)	(6)	(2)
(1)	(1)	(3)	(5)	(4)	(6)	(2)
(3)	(3)	(5)	(1)	(6)	(2)	(4)
(5)	(5)	(1)	(3)	(2)	(4)	(6)
(4)	(4)	(6)	(2)	(1)	(3)	(5)
(6)	(6)	(2)	(4)	(3)	(5)	(1)
(2)	(2)	(4)	(6)	(5)	(1)	(3)

Compare tables 2.3 and 2.4. In the latter case there are no reflective transformations and the cyclic shifts all slope the same way.

Exercises

2.10   Find the two remaining subgroups in table 2.3.

2.11   Find, if possible, other subgroups in the cyclic group of order six (table 2.4) besides the one already mentioned $\{(1), (3), (5)\}$.

2.12   Find a multiplicative group of order 6 among the residue classes modulo 9 and also among the residue classes modulo 14. Arrange the elements in each case so that the multiplication table shows the characteristic diagonal pattern of the cyclic group C_6.
Why is it impossible to find an example of the other group of six elements D_3 amongst the residue classes to any modulus?

2.13   Show that it is not possible to have the same element repeated in any row of a group table, i.e. that it is inconsistent with the group axioms that $a * b = d$ and $a * c = d$ if b and c are two distinct elements of the group.
Is it possible to have a repeated element in any column?

2.14   Experiment with the possibility of finding a group of order 4 other than a cyclic group or Klein's group.

3

SOME WAYS OF ORGANISING A GROUP

3.1 Cyclic groups: regular polygons and star polygons

The number a^3 is óften defined as the product of three factors each
equal to a. Another way of defining it is to say that a^3 is the result of
multiplying the identity element 1 by a three times in succession. This
has the advantage of making sense of the rather baffling expression a°.
To multiply 1 by a zero-times naturally leaves it unchanged suggesting
that $a^\circ = 1$. It is convenient to adopt this notation in the study of
groups and to describe any cyclic group of order n as consisting of
the elements 1, a, a^2, a^3 ... a^{n-1}. For example the non-zero residue
classes (mod 7), may be written 3°, 3^1, 3^2, 3^3, 3^4, 3^5 or otherwise 1, 3, 2,
6, 4, 5. The multiplication table of the elements written in this order
shows the characteristic diagonal pattern of a cyclic group.

The symbolism of a multiplicative group may usefully be adopted
to cover all kinds of cyclic groups. For example the residue classes
0, 2, 4, 6, 8 (mod 10) form an *additive* group of order 5. In this group
we should identify a° as the identity element 0, a as 2, and a^4 as 8 (the
result of adding 2 four times in succession to 0). Again if we consider
the rotation of a square in its own plane as an example of the cyclic
group C_4 with elements a°, a^1, a^2, a^3, then

$a^\circ =$ zero rotation,

$a \ \ =$ a quarter-turn,

$a^2 =$ a half-turn, i.e. 2 quarter-turns in succession.

With this notation it is easy to see why multiplication (mod 7) of the
6 non-zero elements is isomorphic to addition (mod 6). Both sets of

elements may be arranged on clocks in cyclic order which show the connection clearly:

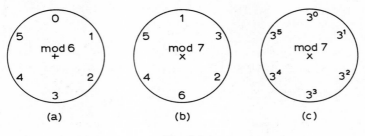

Fig. 3.1

(b) and (c) are identical clocks and it is clear from comparing (a) with (c) that *multiplication* of elements (mod 7) leads to *addition* of the corresponding elements (mod 6). We have in fact a pocket system of logarithms. The complete mapping $0 \longrightarrow 1$, $1 \longrightarrow 3$, $2 \longrightarrow 2$, . . . connecting clock (a) with clocks (b) and (c) could be expressed algebraically in one line as:

$$x \longrightarrow 3^x.$$

Regular polygons are used repeatedly to illustrate cyclic groups. We shall now see that regular polygons and star polygons are of great assistance in understanding the essential structure of congruence classes.

Five points 0, 1, 2, 3, 4, evenly spaced on a circle form a regular pentagon. If alternate points are joined the result is a pentagram.

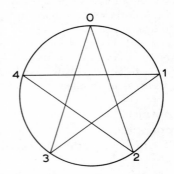

Fig. 3.2

40

Joining every third point leads to the same pentagram in reverse order, and joining every fourth point reverses the original pentagon. The pentagram is an example of a star polygon.

If the basic n-gon has 11 sides then four different star polygons may be inscribed in the circle, beginning respectively 0,2 . . . 0,3 . . . 0,4 . . . and 0,5. . . . On the other hand, if n is a composite number such as 12, the situation is different. There is one star polygon only corresponding to the number 5 (or 7) which is prime to 12. Its points in order form the unicursal path 0,5,10,3,8,1,6,11,4,9,2,7,0. Joining every third point leads to a square 0,3,6,9,0, and the pencil must be lifted twice to bring in the remaining points in two more squares 1, 4, 7, 10, 1, and 2, 5, 8, 11, 2. Alternate points make a pattern of two hexagons, and joining every fourth point gives four equilateral triangles.

These simple facts about polygons are helpful towards understanding some of the properties of structures introduced in Chapters 1 and 2.

(a) Zero divisors (see section 1.15) for the composite modulus 12 are related to the system of squares, hexagons, and triangles just mentioned.

(b) The polygons can be used to illustrate sequences of powers of residues to a prime modulus p. There are $(p-1)$ non-zero residues and $(p-1)$ must be a composite number having at least the factor 2. This means that some but not all of the residues can have their powers arranged on a circle as a polygon or star polygon of order $(p-1)$. To make this point clearer take $p=11$, $(p-1)$ being 10.

First the elements may be arranged on a clock spaced so as to form a regular polygon with powers of 2 at successive vertices.

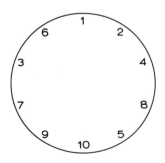

Fig. 3.3

41

This done the clock then illustrates *four* sequences of powers determined by the four regular unicursal paths which include all ten points:

$$1, 2, 4, 8, 5, 10, 9, 7, 3, 6 \text{ (powers of 2)}$$
$$1, 6, 3, 7, 9, 10, 5, 8, 4, 2 \text{ (powers of 6)}$$
$$1, 7, 5, 2, 3, 10, 4, 6, 9, 8 \text{ (powers of 7)}$$
$$1, 8, 9, 6, 4, 10, 3, 2, 5, 7 \text{ (powers of 8)}$$

(c) The polygons are an aid to the discovery of the automorphisms of a cyclic group, mentioned briefly in sections 1.14 and 2.11. As already stated an automorphism is an isomorphism between two arrangements of the elements of a group; and if we choose to regard an automorphism as a one-way mapping of the elements $e, a, b, c, \ldots$, on to the elements $e, a', b', c', \ldots$, then it can be proved, though the proof will be deferred to Chapter 9, that the complete set of automorphisms themselves form a group, the rule of combination being succession.

Let us now consider this statement in relation to the four power series of the residue classes mod 11 set out above. If we name the four automorphisms (1), (2), (3), (4), then mapping (2) may be set out in full as follows:

(2) $1 \longrightarrow 1$　　　　　　$10 \longrightarrow 10$
　　$2 \longrightarrow 6$　　　　　　$9 \longrightarrow 5$
　　$4 \longrightarrow 3$　　　　　　$7 \longrightarrow 8$
　　$8 \longrightarrow 7$　　　　　　$3 \longrightarrow 4$
　　$5 \longrightarrow 9$　　　　　　$6 \longrightarrow 2$

while the first few correspondences of the other three mappings are:

(1) $1 \longrightarrow 1$　(3) $1 \longrightarrow 1$　(4) $1 \longrightarrow 1$
　　$2 \longrightarrow 2$　　　$2 \longrightarrow 7$　　　$2 \longrightarrow 8$
　　$4 \longrightarrow 4$　　　$4 \longrightarrow 5$　　　$4 \longrightarrow 9.$

It may then easily be verified by the reader that the inverse mapping of (2) is

$$1 \longrightarrow 1, \qquad 6 \longrightarrow 2, \qquad 3 \longrightarrow 4, \ldots,$$

and that this when rearranged is exactly the same as (2). However (3) and (4) are not self-inverse. The mapping (3) repeated twice leads to mapping (2) and if repeated three times to mapping (4). This suggests that the four mappings belong to a cyclic group. Set out in detail they may be combined by succession as follows:

42

first mapping

S	(1)	(3)	(2)	(4)
(1)	(1)	(3)	(2)	(4)
(3)	(3)	(2)	(4)	(1)
(2)	(2)	(4)	(1)	(3)
(4)	(4)	(1)	(3)	(2)

second mapping (left of table) — resulting mapping (right of table)

From the table, $(2)S(3)=(4)$.

If we follow the fate of the element 2 under the various mappings we see that

$$2 \longrightarrow 7 \text{ under mapping } 3,$$
$$7 \longrightarrow 8 \text{ under mapping } 2.$$

Hence $2 \longrightarrow 8$ in the resulting mapping which is confirmed by finding that $2 \longrightarrow 8$ under mapping (4).

(d) We are now in a position to find very quickly the automorphisms of a group arising from a prime modulus. Take, for example, $p=17$ and examine the powers of its residues beginning with the smallest:

2: 1, 2, 4, 8, *16*, 15, 13, 9, 1. This sequence fails because 1 recurs too soon, before all the elements have been named.

3: 1, 3, 9, 10, 13, 5, 15, 11, *16*, ... This succeeds because $(p-1)$, i.e. 16, has come at the halfway mark as it should. The clock can now be completed and the automorphisms read off by tracing the various polygons having 16 sides:

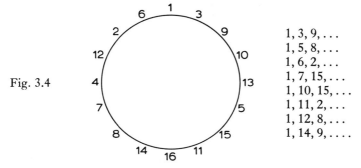

Fig. 3.4

1, 3, 9, ...
1, 5, 8, ...
1, 6, 2, ...
1, 7, 15, ...
1, 10, 15, ...
1, 11, 2, ...
1, 12, 8, ...
1, 14, 9, ...

Any one of these sequences might have been used to name the clock and the others could have been derived just as easily.

43

If the successive powers of any element x form the complete group, x is said to be a *generator* of the group. There are 8 generators in the group just discussed, namely 3, 5, 6, 7, 10, 11, 12, 14. An automorphism of the group may be associated with each generator.

Apart from the special case of the automorphisms in the group of non-zero residues to a prime modulus for multiplication, the star polygons illustrate the automorphisms to be found among any cyclic group containing k elements. For example, if $k=12$, the group might be thought of as residues modulo 12 for addition or as rotations of $0°$, $30°$, $60°$ . . . $330°$ for succession. Generators in these two cases would be 1, 5, 7, 11, and rotations of $30°$, $150°$, $210°$, $330°$, and there would be four automorphisms of the group. If k is a prime number every element except zero qualifies as a generator and there are $k-1$ automorphisms of the group.

There is some opportunity for confusion here with the material discussed in section 3.1(b), so it is worth while summarising the position for residues to a prime modulus p:

(i) There are p residue classes altogether which form a cyclic group for the operation of addition and the group has $p-1$ automorphisms.

(ii) The non-zero residues form a cyclic group of order $p-1$ for the operation of multiplication. If the number $p-1$ has n co-primes each less than p, the group has n automorphisms.

3.2 Primitive roots and quadratic residues

In connection with congruences another name for a generator when multiplication is the rule of combination, is 'primitive root'. It has been proved that a primitive root exists for every prime number p, but the proof is outside the scope of this book. As already seen, if one primitive root is known then others may quickly be derived and they play an important part in congruences, having many uses.

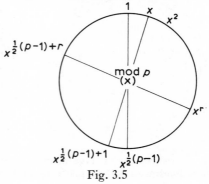

Fig. 3.5

44

A more general clock can now be drawn in terms of a primitive root x. The fact already noticed that $p-1$ is the element opposite 1 may be expressed thus:

$$x^{(p-1)/2} = p - 1 \;(\text{mod } p),$$
$$x^{(p-1)/2} = -1,$$
$$1 + x^{(p-1)/2} = 0. \tag{3.1}$$

This enables us to generalise another interesting feature which may have been noticed on the multiplication clocks used for various prime numbers. Elements at opposite ends of a diameter always seem to be complementary (modulo p). In the 17-clock 14 is opposite 3 and 8 is opposite 9. In general, if x is a primitive root, numbers at opposite ends of a diameter must be of the form x^r, $x^{r+(p-1)/2}$ and their sum is $x^r(1 + x^{r(p-1)/2})$. Since $1 + x^{(p-1)/2} = 0$ from equation (3.1), the sum is zero and the two numbers are complementary.

It is sometimes of great importance to know which of the residues to a prime modulus p are squares. If $p = 11$, then the squares of the residues 0—10 in order are 0, 1, 4, 9, 5, 3, 3, 5, 9, 4, 1. Only the residues 0, 1, 4, 9, 5, 3, are perfect squares and they are called quadratic residues, the others quadratic non-residues. For obvious reasons, on an 11-clock (Fig. 3.3) the non-zero quadratic residues necessarily appear at the even nodes, and in general once a primitive root has been discovered its even powers are all quadratic residues.

As an example of their use consider the quadratic congruence $x^2 - 6x + 3 \equiv 0 \;(\text{mod } 11)$. The roots if they exist are

$$x \equiv \frac{6 \pm \sqrt{2}}{2}.$$

Since 2 is a quadratic non-residue the roots do not exist. On the other hand, the roots of $x^2 - 6x + 5 \equiv 0 \;(\text{mod } 11)$ are

$$x \equiv \frac{6 \pm \sqrt{5}}{2},$$

and since $5 \equiv 4^2$ or 7^2 this gives us two roots to the congruence $x \equiv 5$ or 1.

Another example of the use of quadratic residues will be found in Chapter 7.

Exercises

3.1 Complete the sequence $S_1 = 1, 2, 4, \ldots$ of residue classes modulo 19 expressed as powers of 2.

3.2 Which are the quadratic residues to this modulus?

3.3   Only one of the three equations following can be solved in arithmetic modulo 19. Which one is it and what are its two roots?

$$x^2 + x + 2 = 0,$$
$$x^2 + x + 3 = 0,$$
$$x^2 + 3x + 1 = 0.$$

3.4   Write S_1 in column 1 and in column 2 put S_2 derived from S_1 by taking every fifth number. Derive S_3 from S_2 in the same way and continue if necessary with S_4, etc., until you have decided how many automorphisms the multiplicative group of non-zero residues possesses and have identified the group of these automorphisms.

3.3   The order of an element; cyclic subgroups

In analysing a group one of the most fruitful enquiries is into the *orde* of the various elements. To determine the order of an element x we observe the point at which the identity element recurs in the sequence $1, x, x^2, \ldots$ Suppose that $x^3 = 1$. This indicates the presence of a subgroup with three elements and a multiplication table as follows:

	1	x	x^2
1	1	x	x^2
x	x	x^2	1
x^2	x^2	1	x

If the group is large further enquiry might reveal the existence of an element y of order 6 such that $y^2 = x$, in which case there would also be a larger cyclic subgroup with elements $1, y, x, y^3, x^2, y^5$, or otherwise expressed $1, y, y^2, y^3, y^4, y^5$.

Determination of the order of the various elements of a group also gives some insight into its nature and may possibly suggest a good geometrical representation. For example, the existence of many elements of orders 3 and 4 would suggest the cube and the octahedron as likely geometrical models to be related to the group, because they have rotational symmetry of orders 3 and 4.

3.4 Subgroups of D₄

The multiplication table of a group may be presented in a random order or it may be organised to show one or more of the regular patterns which underlie its structure. It is profitable to look at a particular group of sufficiently high order to possess interesting features and to organise it in various ways using first one and then another of its subgroups as a basic unit. The dihedral group of order 8, sometimes called the octic group and designated D_4, may be chosen for this purpose. A common representation of D_4 is the set of eight geometrical transformations which permute 8 points symmetrically disposed on a circle to form a Maltese-cross pattern (fig. 3.6). Another representation was mentioned in section 2.4.

The table can be written down first on the pattern common to dihedral groups, following that already set out for D_3 in section 2.5.

Table 3.1

points

		1	2	3	4	5	6	7	8
	(1)	1	2	3	4	5	6	7	8
	(2)	2	3	4	1	8	5	6	7
	(3)	3	4	1	2	7	8	5	6
transformations	(4)	4	1	2	3	6	7	8	5
	(5)	5	6	7	8	1	2	3	4
	(6)	6	7	8	5	4	1	2	3
	(7)	7	8	5	6	3	4	1	2
	(8)	8	5	6	7	2	3	4	1

As already explained in section 2.2, this table leads immediately to a table of transformations. From this point in the book any such table will be printed without brackets and the statement 2S5=8 will be taken to mean that transformation (2) succeeding transformation (5) is equivalent to transformation (8).

47

The points on the Maltese cross can now be named to conform with the table.

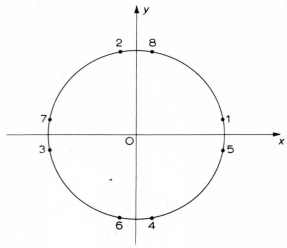

Fig. 3.6

If the point in the first quadrant is chosen to be 1, the cyclic subgroup in the top left-hand compartment suggests rotations of 90° and gives meaning to operations (1) to (4):

        (1) Stand still.
        (2) Rotate 90° anti-clockwise, (sending point 1 to 2).
        (3) Rotate 180° anti-clockwise, (sending 1 to 3).
        (4) Rotate 270° anti-clockwise, (sending 1 to 4).

The point 5 is now chosen, adjacent to 1, and it is observed in the table that under transformation (2) the points 5, 6, 7, 8, have a cyclic shift opposite in direction to that of 1, 2, 3, 4, so the remaining points are named in clockwise order. Transformations (3) and (4) may be checked for consistency with these names. Transformation (5) produces interchanges $1 \longleftrightarrow 5$, $2 \longleftrightarrow 6$, $3 \longleftrightarrow 7$, $4 \longleftrightarrow 8$ and is clearly a reflection in the x-axis. The list of transformations may now be completed:

        (5) Reflect in x-axis.
        (6) Reflect in $y = -x$.
        (7) Reflect in y-axis.
        (8) Reflect in $y = x$.

48

The figure contains rectangles which suggests the possibility of Klein's four-group as a sub-group. Two rectangles contain the point 1 and investigation shows that (1), (3), (5), (7) and (1), (3), (6), (8), are sub-groups. The table can now be rearranged on one of these:

Table 3.2

S	1	3	5	7	2	4	6	8
1	1	3	5	7	2	4	6	8
3	3	1	7	5	4	2	8	6
5	5	7	1	3	6	8	2	4
7	7	5	3	1	8	6	4	2
2	2	4	8	6	3	1	5	7
4	4	2	6	8	1	3	7	5
6	6	8	4	2	7	5	1	3
8	8	6	2	4	5	7	3	1

The following points may be noted:

(i) The elements of the subgroup {1, 3, 5, 7} are repeated in the bottom right-hand corner but in a different order; the coset {2, 4, 6, 8} is also repeated in a different order elsewhere.

(ii) Every compartment regarded as a set of elements exhibits the characteristic of a Klein's group, namely that of interchanging in pairs between any two lines or columns.

(iii) The compartments themselves have the pattern of a group, in this case of order 2:

	A	B
A	A	B
B	B	A

where A stands for the set {1, 3, 5, 7} and B for the set {2, 4, 6, 8}.

(iv) In this group the set A of elements in the subgroup plays the part of the identity element.

3.5   Right and left cosets, normal subgroups

The group also contains subgroups of order two and can be organised in cosets of two elements each, using one or other subgroup as a basis. If we multiply the subgroup {1, 5} on the right by any element we get a *right coset*. For example {1, 5}×2 leads to the right coset {2, 6} and elements 3 and 4 treated in the same way yield two more, {3, 7} and {4, 8}. A *left coset* comes from multiplying the subgroup on the left by any element, the cosets being {1, 5} {2, 8} {3, 7} and {4, 6}. We see that in this case left and right cosets are not the same.

Another subgroup {1, 3} however produces left and right cosets which are exactly the same, namely {1, 3}, {2, 4}, {5, 7}, and {6, 8}. This marks an important difference between the two subgroups. It is instructive to set out the table for the group using first {1, 5} and then {1, 3} as an organiser.

Table 3.3

	1	5	2	6	3	7	4	8
1	1	5	2	6	3	7	4	8
5	5	1	6	2	7	3	8	4
2	2	8	3	5	4	6	1	7
6	6	4	7	1	8	2	5	3
3	3	7	4	8	1	5	2	6
7	7	3	8	4	5	1	6	2
4	4	6	1	7	2	8	3	5
8	8	2	5	3	6	4	7	1

Table 3.4

	1	3	2	4	5	7	6	8
1	1	3	2	4	5	7	6	8
3	3	1	4	2	7	5	8	6
2	2	4	3	1	8	6	5	7
4	4	2	1	3	6	8	7	5
5	5	7	6	8	1	3	2	4
7	7	5	8	6	3	1	4	2
6	6	8	7	5	4	2	1	3
8	8	6	5	7	2	4	3	1

Comparison of the two tables for the second and fourth rows of compartments shows that {1, 3} is a much better organiser than {1, 5}. Whereas in table 3.3 some compartments contain two and others four distinct elements, each compartment in table 3.4 contains exactly two distinct elements, the pair in each case being the same as a coset pair.

This enables us to set out a new table showing the pattern of compartments:

Table 3.5

	A	B	C	D
A	A	B	C	D
B	B	A	D	C
C	C	D	A	B
D	D	C	B	A

where A$=\{1,3\}$, B$=\{2, 4\}$, C$=\{5, 7\}$, and D$=\{6, 8\}$. The table may be recognised as having the same structure as Klein's group.

A good organiser such as A is called a *normal subgroup* and it plays the part of the identity element in the structure whose elements are A, B, C, D which is called its quotient group. Normal subgroups, which will be discussed in greater detail with more precision in chapter 6 are of fundamental importance in the theory of groups. One way of expressing their distinguishing property is to say that the right and left cosets of a normal subgroup are the same. Another way is to say that its elements taken as a whole, not individually, commute with every other element. For example in table 3.2 it can be seen that $\{1, 3, 5, 7\} \times 2 = \{2, 4, 6, 8\}$ while $2 \times \{1, 3, 5, 7\} = \{2, 4, 8, 6\}$ but the element 5 individually does not commute with either 6 or 8.

Again in table 3.4 $\{1, 3\}$ commutes with $\{2, 4\}$ but in table 3.3 $\{1, 5\}$ does not commute with $\{2, 6\}$ thus showing that $\{1, 5\}$ is not a normal subgroup. It should be noted that the whole group and the single element e comply with the conditions for a normal subgroup.

The process of breaking a group down into smaller elements by means of normal subgroups is something like the process of factorisation of a composite number in ordinary arithmetic, and for this reason the quotient group is sometimes called the factor group. However, normal subgroups and quotient groups do not always exist when one would expect them and a notable exception is the group of 60 rotations of an icosahedron. This group has no normal sub-groups other than the trivial cases of the whole group and the group consisting only of e, and is therefore called a simple group. All cyclic groups of prime order are examples of simple groups.

In table 3.3 it is worth noticing that the right cosets by definition appear in the columns of the first row of compartments, while the left cosets come in the rows of the first column.

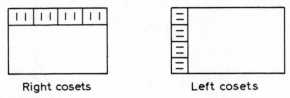

Right cosets   Left cosets

Fig. 3.7

Exercise

3.5   This is the 'multiplication' table of a group:

Table 3.6

	a	b	c	d	e	f	g	h	j	k	l	m
a	c	m	l	g	k	d	f	e	b	h	a	j
b	e	l	d	c	a	m	h	g	k	j	b	f
c	l	j	a	f	h	g	d	k	m	e	c	b
d	b	k	e	m	g	h	c	j	f	a	d	l
e	d	f	b	h	j	c	m	a	l	g	e	k
f	j	e	h	b	d	k	a	m	g	l	f	c
g	m	h	k	j	f	e	l	b	d	c	g	a
h	f	g	j	k	m	a	b	l	c	d	h	e
j	h	c	f	a	l	b	k	d	e	m	j	g
k	g	d	m	e	b	l	j	c	a	f	k	h
l	a	b	c	d	e	f	g	h	j	k	l	m
m	k	a	g	l	c	j	e	f	h	b	m	d

(a) What is the identity element?

(b) An element b such that $b^n = e$ (the identity element) is said to be of

52

order n. The element c is of order 3. Find the orders of each of the other elements in this group.

(c) Name two subgroups of order 3.

(d) Find the inverses of m, f, j, l.

(e) Is the group commutative?

(f) Verify the associative property for a, b, c and g, k, g.

(g) Show that b, h, l are part of a subgroup of order 4 and name the other element. Is the subgroup cyclic or Klein's?

Multiply each element in the subgroup by c, writing c on the left and so forming a left coset. Find another left coset. Also find the right cosets of the subgroup. Is it a normal subgroup? If so what is its quotient group?

(h) Find all the right and left cosets of the subgroup l, f, k. Is it a normal subgroup?

(i) Could either the rotations of a regular hexagon or those of a regular tetrahedron be a representation of this group?

3.6   Geometrical significance of cosets

An example of the relationship which exists between a subgroup and its other cosets may be clearly seen in the geometric representation of D_4 (fig. 3.6). The cyclic subgroup $\{1, 2, 3, 4\}$ relates 4 points at the corners of a square and its coset $\{2, 4, 6, 8\}$ also makes a square pattern. There is a rectangle associated with $\{1, 3, 5, 7\}$ and another rectangle with its coset. Then $\{1, 3\}$ names a diameter and so does each of its cosets; on the other hand $\{1, 5\}$ which is not a normal subgroup determines a chord which, with its right cosets leads to a set of parallel chords, and its left cosets a set of chords which are alternate sides of an octagon, two rather weaker relationships though still having some regularity.

The group D_6 is likely to have an abundance of subgroups and the reader may like to write down its table on the same pattern as for D_4 and then rearrange it using a variety of subgroups as bases in turn. Six pairs of points regularly placed on a circle would provide a suitable illustration, containing sets of points which would suggest likely subgroups.

3.7 Organisation by cosets of finite rings

It was mentioned in section 1.15 that when n is a composite integer, Z_n forms a ring. Formal definition of a ring is deferred till chapter 7 but we already know a good deal about Z_n. Its n elements, the residue classes modulo n, form an additive group, but the $(n-1)$ non-zero elements do *not* form a group for multiplication. Let us consider Z_{15}. The multiples of 5, namely $\{0, 5, 10\}$ form a normal sub-group for addition and the whole ring can be organised in a very striking manner if we use this sub-group as a basis. The reader is invited to write out the multiplication table for the 15 elements using the following order:

$$0, 10, 5; \quad 6, 1, 11; \quad 12, 7, 2; \quad 9, 4, 14; \quad 3, 13, 8.$$

The structure revealed by this exercise will explain itself more eloquently than a description in words. It is also instructive to re-write the table putting the elements in a different order using multiples of 3 as an organiser:

$$0, 6, 12, 9, 3; \quad 10, 1, 7, 4, 13; \quad 5, 11, 2, 14, 8.$$

The remarkable thing is that the additive sub-group and its cosets are components of both the addition and the multiplication tables.

Exercise

3.6 Explore the structure of Z_{10}, Z_{12}, and Z_9.

(a) Try to forecast the possible structure of the multiplication table in each case.

(b) Find a normal subgroup for addition and its cosets. Then write out the multiplication table, using the same cosets in an appropriate order. Compare your results with your forecast in each case.

3.8 Generators and relations

A cyclic group of order n can be expressed in terms of one generator only, as has already been noticed, the elements being $1, a, a^2, \ldots, a^{n-1}$. Other groups need more than one generator to define them completely, together with some relations between the generators. For example Klein's group can be developed from two generators a, b and the relations $(1) a^2 = b^2 = 1$, $(2) ab = ba$. The full table can be obtained from these relations in terms of a, b and 1.

$*$	1	a	b	ab
1	1	a	b	ab
a	a	1	ab	b
b	b	ab	1	a
ab	ab	b	a	1

The last product is obtained as follows:

$$(ab)(ab) = a(ba)b, \quad \text{associative property,}$$
$$= a(ab)b, \quad \text{relation (2),}$$
$$= a^2b^2, \quad \text{associative property,}$$
$$= 1 \times 1, \quad \text{relation (1),}$$
$$= 1, \quad \text{identity property.}$$

Generators and relations are a compact and convenient way of specifying a group, though the specification does not display such complete and detailed information as the full table.

Exercise

3.7 Develop groups of order 8 from the following generators and relations:

(a) $a^4 = b^2 = 1$, $ab = ba$.

The 8 elements may be called 1, a, a^2, a^3, b, ab, a^2b, a^3b.
Is this a commutative group?

(b) $a^4 = b^2 = 1$, $ba = a^3b$.

3.8 Suggest likely generators and relations for (a) the dihedral group of order 6, D_3, (b) the dihedral group of order 10, D_5.

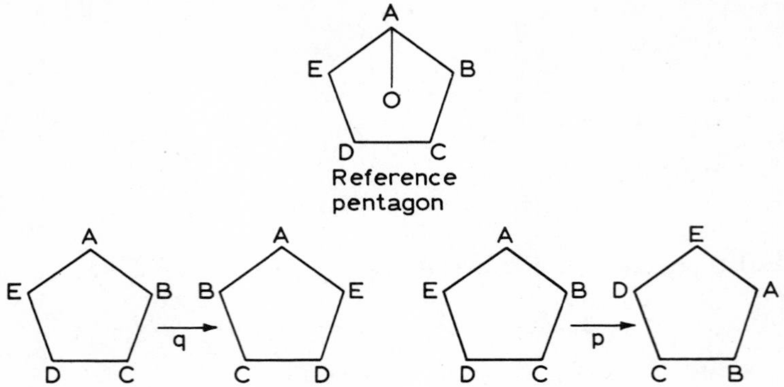

Reference
pentagon

Fig. 3.8

The figures above show p as a clockwise rotation of 72° about 0 applied to the reference pentagon and q as a half-turn about OA. pq=p succeeding q. The identity transformation is e.

(a) Express a half-turn about OB in terms of p and q.

(b) Show that $qp=p^4q$.

(c) Hence or otherwise show that

$$\begin{array}{ll} \text{(i)} & pqp=q, \\ \text{(ii)} & (qp^2)^2=e, \\ \text{(iii)} & qp^3q=p^2. \end{array}$$

3.9 Transforms and automorphisms

In any group it is often found useful to consider the element $a^{-1}ba$ where a and b are any elements and a^{-1} is the inverse of a such that $aa^{-1}=a^{-1}a=e$, the identity element. The derived element $a^{-1}ba$ is called the transform of b by a. In D_4 (table 3.5) 2 and 4 are inverse elements and if we assign the fixed value 2 to the symbol a while varying b we obtain a complete set of transforms:

$$\begin{array}{c} \text{For example, if } b=5 \\ 4\times5=6 \\ 6\times2=7, \\ \text{and hence } a^{-1}ba=7. \end{array}$$

Table 3.7

a^{-1}	b	a	$a^{-1}ba$
4	1	2	1
4	2	2	2
4	3	2	3
4	4	2	4
4	5	2	7
4	6	2	8
4	7	2	5
4	8	2	6

The mapping defined by $b \longrightarrow (a^{-1}ba)$ is always an automorphism of the group: in every case the product of the transforms is the transform of the product. To verify this statement for a particular case take elements 3 and 5 from D_4.

$$3 \times 5 = 7, \text{ from table 3.4.}$$

Transform of $3 \times$ transform of $5 = 3 \times 7$, from table 3.7,

$$= 5, \text{ from table 3.4,}$$

$$= \text{transform of 7.}$$

Verification in every case even for this one group would be extremely tedious but the general case for all groups can be proved, using the group properties.

(Transform of b) $\times$ (transform of c) $= (a^{-1}ba) \times (a^{-1}ca)$,

$$= (a^{-1}b) \times (a\, a^{-1}) \times (ca), \text{ associative,}$$

$$= (a^{-1}b) \times 1 \times (ca), \text{ inverse,}$$

$$= (a^{-1}b) \times (ca), \text{ identity,}$$

$$= a^{-1} \times (bc) \times a, \text{ associative,}$$

$$= \text{transform of } bc.$$

c

Hence the system $(a, b, c, \ldots)$ is isomorphic with its set of transforms, and a method is at hand for finding some but not necessarily all of the automorphisms of the system. The full table of transforms of b by a for all a and b in D_4 is set out below:

Table 3.8

		1	2	3	4	5	6	7	8
	1	1	1	1	1	1	1	1	1
	2	2	2	2	2	4	4	4	4
	3	3	3	3	3	3	3	3	3
	4	4	4	4	4	2	2	2	2
b	5	5	7	5	7	5	7	5	7
	6	6	8	6	8	8	6	8	6
	7	7	5	7	5	7	5	7	5
	8	8	6	8	6	6	8	6	8

The four automorphisms defined by the columns of this table (each being repeated twice) do not exhaust the possibilities. They are called inner automorphisms, a term reserved for those obtained by the transform method. Any others are called outer automorphisms. For the group D_4 which we are considering, there happen to be four others. Although they do not arise directly from the transform method they may be inferred from studying the pattern of the inner automorphisms. In table 3.8 the cyclic order of the subsets 1, 2, 3, 4, and 5, 6, 7, 8, is never upset. Using this clue four other columns may be added to Table 3.8 and the validity of the resulting automorphisms may be verified in particular cases.

1	1	1	1	1
2	2	2	4	4
3	3	3	3	3
4	4	4	2	2
5	6	8	6	8
6	7	5	5	7
7	8	6	8	6
8	5	7	7	5

Although now we cannot use the word 'transform' the language of mappings will serve instead and our test will be that 'the product of the images is the image of the product'.

It should be mentioned before leaving this topic that commutative groups can have no inner automorphisms other than the identity automorphism, since

$$a^{-1}(ba)=a^{-1}(ab),$$
$$=(a^{-1}a)\,b,$$
$$=b,$$

i.e. every element is its own transform for each a and b.

As an example of the use of transforms consider a line $y=mx$ making an angle α with the x—axis.

Fig. 3.9

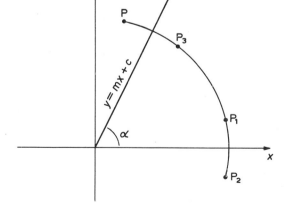

59

Define two transformations a and b as follows:

> $a =$ a clockwise rotation through angle α about O,
> $b =$ reflection in the x—axis.

Then $a^{-1}ba = a$ followed by b followed by a^{-1}.

The effect on P, any point in the plane, of these three transformations in succession is as indicated in the diagram, $P \longrightarrow P_1 \longrightarrow P_2 \longrightarrow P_3$ and it is clear that the resulting transformation or mapping is $P \longrightarrow P_3$ and that this is the same as reflection in the line $y = mx$. For those familiar with matrix algebra it will be clear that as a, b, a^{-1} are standard transformations this is a convenient method of expressing the transformation $P \longrightarrow P_3$ as the product of three standard matrices.

3.10 Conjugate elements

Table 3.8 also reveals the existence of classes among the elements. Looking along the rows it is seen that 2 and 4 have an affinity: each transforms sometimes into the other and never into any other element. They are said to be conjugate elements and together to form a conjugacy class. Other classes are $\{5, 7\}$ and $\{6, 8\}$.

Conjugate elements have the same part to play in the structure of the group. This is best seen perhaps from the geometrical representation in section 3.4 where it was agreed that (2) and (4) could be interpreted as rotations of 90°. Both operators are of order 4 and a rotation of 90° in one sense is the same as 270° in the opposite sense. If we had thought of clockwise rotation being positive we should probably have called (4) by the name of (2) and vice versa. In the same way (5) as an operator is akin to (7) and (6) to (8); but there is no operator akin to a rotation of 180° and so no element transforms into 3. Because 3 always transforms into 3 it is called a *self-conjugate element*.

Exercises

3.10 List the inverses of all elements in the group specified by table 3.6.

3.11 (a) Find the classes of conjugate elements in this group. Verify that elements in the same class are of the same order.

(b) Prove that this is a general property for all groups.

3.12 In table 3.6 list all the elements which commute with the element c (including c itself). Verify that these elements form a subgroup of the group. The subgroup is called the normaliser of c.

3.13 (a) If in any group a commutes with both b and c prove that it also commutes with (bc).

(b) Show that in any group the set of elements which commutes with a (including a itself) forms a subgroup of the group.

3.14 Refer to question 3.12 and write down any right coset of the subgroup obtained (multiplying each element of the subgroup on the right by any other element of your own choice). Verify that each element of the coset transforms c alike. Choose a second coset and investigate the transforming effect on c of each of its elements.

3.15 Repeat questions 3.12 and 3.14 for the element h.

3.11 Another group of order 8, $C_2 \times C_2 \times C_2$

A set of three elements $\{a, b, c\}$ has six proper subsets $\{a, b\}$, $\{b, c\}$, $\{c, a\}$, $\{a\}$, $\{b\}$, $\{c\}$. If the two improper subsets $\{\ \ \}$ and $\{a, b, c\}$ are included, the subsets can be exhibited as a group under the operation symmetric difference for which the conventional symbol is Δ. For two sets A, B the symmetric difference $A \, \Delta B$ is defined to be the set of elements which are in either A or B but not in both:

$$\{a, b\} \, \Delta \{b, c\} = \{a, c\},$$
$$\{b\} \, \Delta \{\quad\} = \{b\},$$
$$\{a\,b\,c\} \, \Delta \{c\} = \{a\,b\}.$$

The empty set acts as identity element and the group behaves very much like Klein's group: every pair of lines or columns is related by interchanging pairs. All its subgroups are Klein's groups or groups of order 2. Each of the latter has a Klein's group as its quotient group.

The same group would arise from the set of polynomials in x of second degree or less whose coefficients obey the rules for modulus 2. The possible polynomials are $0, 1, x, x+1, x^2, x^2+1, x^2+x, x^2+x+1$. Any two members of the set when added give a third member, e.g. $(x^2+1)+(x^2+x)=x+1$, since $2x^2=0 \pmod 2$. If these elements are set out in a table for the operation of addition the first three lines would be:

$+$	0	1	x	$x+1$	x^2	x^2+1	x^2+x	x^2+x+1
0	0	1	x	$x+1$	x^2	x^2+1	x^2+x	x^2+x+1
1	1	0	$x+1$	x	x^2+1	x^2	x^2+x+1	x^2+x
x	x	$x+1$	0	1	x^2+x	x^2+x+1	x^2	x^2+1

$\{0, 1\}$ is a normal subgroup; the other elements are arranged according to its cosets. There is enough of the table to indicate its general pattern, interchanging in pairs between any two lines.

The group is commutative and so has no inner automorphisms except the identity automorphism, which is trivial, but there are a number of outer automorphisms. One could be obtained by interchanging the rôles of x and 1 leading to the following mapping:

$$0 \longrightarrow 0 \qquad\qquad x^2 \longrightarrow x^2$$
$$1 \longrightarrow x \qquad\qquad x^2+1 \longrightarrow x^2+x$$
$$x \longrightarrow 1 \qquad\qquad x^2+x \longrightarrow x^2+1$$
$$x+1 \longrightarrow x+1 \qquad x^2+x+1 \longrightarrow x^2+x+1$$

Similarly x and x^2, or x^2 and 1 could be interchanged and another way of obtaining an automorphism would be through a cyclic interchange $x \longrightarrow 1, 1 \longrightarrow x^2, x^2 \longrightarrow x$.

The group is designated $C_2 \times C_2 \times C_2$ and has no cycles of order greater than 2 among its subgroups. It requires 3 separate generators to develop all its elements. A geometrical representation will be found in section 8.6 and another in 11.2.

3.12 The group $C_4 \times C_2$

	a	b	c	d	e	f	g	h
a	d	h	b	e	f	a	c	g
b	h	d	a	g	c	b	f	e
c	b	a	f	h	g	c	e	d
d	e	g	h	f	a	d	b	c
e	f	c	g	a	d	e	h	b
f	a	b	c	d	e	f	g	h
g	c	f	e	b	h	g	d	a
h	g	e	d	c	b	h	a	f

Here the table is in scrambled form. How can it be reduced to order?

(a) First its identity element can be recognised as f.

(b) Next we can obtain powers of its first element a: $a^2=d$, $a^3=ad=e$, $a^4=ae=f$. Hence $\{f, a, d, e\}$ is a cyclic subgroup of order 4.

(c) Choosing any other element such as b the coset is $\{f,\,a,\,d,\,e\}\times b$, i.e. $\{b,\,h,\,g,\,c\}$.

The table can now be organised using $\{f,\,a,\,d,\,e\}$ as a basis:

Table 3.9

✳	f	a	d	e	b	h	g	c
f	f	a	d	e	b	h	g	c
a	a	d	e	f	h	g	c	b
d	d	e	f	a	g	c	b	h
e	e	f	a	d	c	b	h	g
b	b	h	g	e	d	e	f	a
h	h	g	c	b	e	f	a	d
g	g	c	b	h	f	a	d	e
c	c	b	h	g	a	d	e	f

Is there a geometrical representation? The cyclic subgroup suggests rotation of 90° as in D_4, but this time the coset has the same slope as the subgroup; there is no suggestion yet of reflection. So far a possible configuration would be 8 points spaced evenly on a circle in the order $f,\,b,\,a,\,h,\,d,\,g,\,e,\,c$. However, this is not consistent with operation (b) which sends f to b, b to d, d to g, g to f, and also produces the cycle $a\longrightarrow h\longrightarrow e\longrightarrow c$. A more illuminating operation is (h) which produces direct interchanges $f\longleftrightarrow h$, $d\longleftrightarrow c$, $a\longleftrightarrow g$, $e\longleftrightarrow b$.

This suggests reflection, but in what mirror? We have to avoid reversing the order of the cycle $b,\,h,\,g,\,c$. If we consider the rotation to be about a vertical axis then the difficulty can be resolved by choosing a horizontal mirror. As $f,\,a,\,d,\,e$ revolve in a horizontal plane the points $b,\,h,\,g,\,c$ revolve in a parallel plane and reflection in the mirror produces the required interchanges. The position is best understood by means of a diagram.

Fig. 3.10

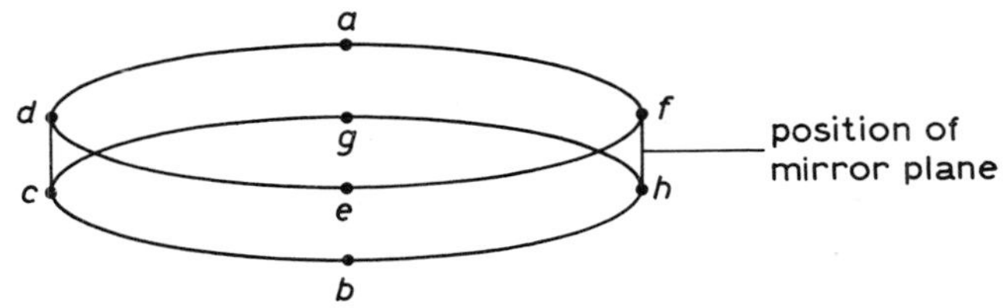

position of mirror plane

63

Operations (g), (c), and (b) fall conveniently into line and the operations may now be listed in full:

- (f) stand still.
- (a) rotate 90° anti-clockwise.
- (d) rotate 180° anti-clockwise.
- (e) rotate 270° anti-clockwise . . . (1).
- (h) reflect in horizontal mirror.
- (g) reflect in horizontal mirror and rotate 90°.
- (c) reflect in horizontal mirror and rotate 180°.
- (b) reflect in horizontal mirror and rotate 260°.

This group appears in chemistry as a crystallographic group and its conventional diagram is given below.

Fig. 3.11

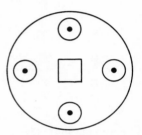

The square at the centre indicates rotational symmetry of order 4 about a vertical axis and the symbol ⊙ indicates an object and its image in a horizontal mirror. A single object, represented by ●, if successively rotated and/or reflected according to the 8 instructions above occupies each of the 8 positions in turn. Operations (g), (c), and (b) are called roto-reflections and are somewhat akin to glide-reflections in 2-space.

Exercises

3.16 Which group is defined by the generators a, b, c and the relations $a^2=b^2=c^2=e$, $ab=ba$, $bc=cb$, $ca=ab$?

3.17 Develop the group of order 8 defined by the generators a, b and the relations $a^4=e$, $b^2=a^2$, $ba=a^3b$. (This group which has not been described in the text is called the quaternion group.) Find the order of each of its elements.

3.18 There are five groups of order 8. Set out a table showing the number of elements of order 1, 2, 4, 8 respectively for each group.

3.19 Suggest suitable generators and relations for the group formed from the subsets of the set $\{a, b, c, d\}$ combined by symmetric difference.

3.20   Find sets of 8 elements which form multiplication groups among the residue classes modulo 15, 16, 17, 24 respectively. Identify the group in each case.

3.13   Movements in square-dances

The groups of transformations of 8 elements may be compared with the connecting figures interpolated between the individual movements of a square-dance for four couples. Here are some of the calls:

Honour and circle left: bow to partner, bow to contrary, join hands in a ring and circle left round to original places.

Grand chain: men and women weave in and out in a circular track, moving in opposite directions.

Promenade: the man takes the woman with crossed hands and leads her round in a counter-clockwise direction to original places.

These are the basic movements for C_8, D_4, and $C_4 \times C_2$. The caller often varies the connecting figure by inserting an extra flourish half-way round the circle:

Balance and honour: bow to each other.

Turn: man takes partner with crossed hands and turns her.

Circle left a quarter way round, balance and honour and circle home would suggest operation (2) of C_8 followed by its inverse operation (6).

Promenade your partner half-way round, balance and turn and promenade home. This command suggests that operation (c) in $C_4 \times C_2$ should be repeated twice, except that reflection in a horizontal mirror has been replaced by a half-turn in a horizontal plane: the first (c)=promenade and half-turn, the second=half-turn and promenade, which produces an equivalent effect. Since (c) is self-inverse this is again a case of a movement and its inverse restoring the original position, expressed by the caller as 'returning home'.

In the initial position men are on the inside and women are on the outside. Exactly half way through the promenade, in the middle of the 'turn', operation (c) has been completed. At this instant the men are on the outside of the circle.

Fig. 3.12

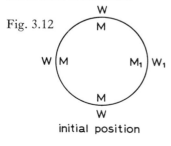

initial position

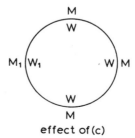

effect of (c)

4

FIBONACCI SEQUENCES

4.1 The Fibonacci sequence

Many readers will be familiar with the famous sequence

$$1, 1, 2, 3, 5, 8, 13, \ldots$$

in which each term after the second is obtained by adding the previous two terms. It was known at least as early as 1226 when Leonardo Fibonacci of Pisa propounded his problem concerning the proliferation of a hypothetical pair of rabbits: each pair of newborn rabbits is assumed to bear its first pair two months later and thereafter to bear one pair a month. (See N. N. Vorob'ev, *Fibonacci Numbers*, Popular Lectures in Mathematics, Vol. 2).

If we adopt a convenient code the development of the rabbit colony can easily be set out:

$$N = \text{a newborn pair,}$$
$$Y = \text{a young pair (one month old),}$$
$$MN = \text{a mature pair with new born pair}$$
$$\text{(two months or older).}$$

Results are as follows:

Time	Description	Pairs in colony
initially	N	1
after 1 month	Y	1
after 2 months	MN	2
after 3 months	MN Y	3
after 4 months	MN Y MN	5
after 5 months	MN Y MN MN Y	8

To continue the sequence indefinitely requires one more assumption – namely that all rabbits are immortal!

66

4.2 The ratio of two consecutive terms

Pairs of consecutive numbers (1,1), (1,2), (2,3), . . . can be plotted on the Cartesian plane. After an irregular start the points appear to be on a straight line or nearly so. If each point is joined to the origin the lines soon become indistinguishable and the collective gradient, if such a loose expression may be permitted, gives a result when measured something like 1.62.

The calculated values for 3/2, 5/3, 8/5 . . . to three places of decimals are: 1·5, 1·667, 1·600, 1·625, 1·616, 1·619, 1·618, . . . They appear to be converging towards some definite value and it may be noticed that alternate numbers 1·5, 1·600, 1·616 . . . continually increase while the others continually diminish, i.e. that the line of best fit threads its way among the plotted points which lie alternately above and below it.

4.3 Other sequences formed from the same rule

The original Fibonacci sequence began with the terms 1,1. Other such sequences may be formed using a different pair of initial terms, such as 1, 3, leading to

$$1, 3, 4, 7, 11, 18, \ldots$$

Pairs of consecutive terms may be plotted as before and, if so, the points are found to lie close to the same line of best fit as for the Fibonacci pairs. It is most instructive to start with an unlikely initial pair such as 2, 19 and build up a sequence 2, 19, 21, 40, . . . Although the points specified by contiguous pairs fluctuate erratically at first it is only necessary to carry the sequence to 10 or 12 terms to find out that the points are inexorably drawn to the same line of best fit as before: the ratio 688/425 for instance gives 1·617.

Empirically it looks as if any pair of initial terms, however far apart, if continued in the Fibonacci way will yield plotted points converging on a definite straight line through the origin. It will be shown later on in this chapter that there is such a line and that its gradient is an irrational number which we shall determine.

It is necessary at this point to clear up a possible ambiguity as to the exact meaning of a 'Fibonacci sequence'. In this book all sequences obtained by adding two consecutive terms to get the next one will qualify for the title. Nevertheless 'the Fibonacci sequence' will refer only to the particular sequence 1, 1, 2, 3 . . . and the members of this particular sequence will be called Fibonacci numbers.

67

4.4 Golden section

The Greeks discovered a ruler-and-compasses method for dividing a line-segment in 'golden section'. The problem proposed was to find a point C on AB such that $AC^2 = AB.BC$.

Fig. 4.1

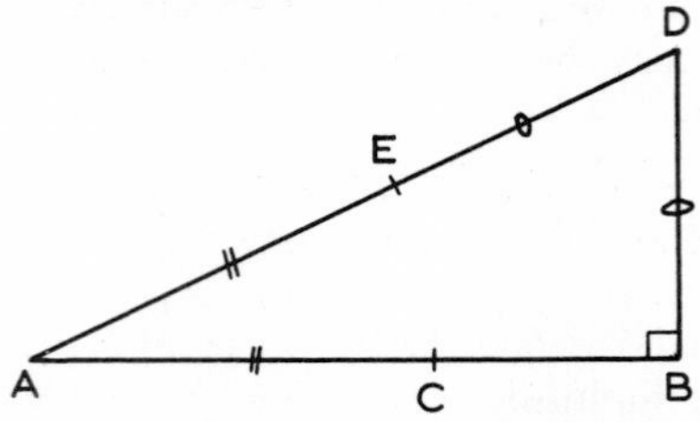

It will be recalled that the method is to erect a perpendicular BD at B, equal to half AB, then to join AD and cut off $DE = DB$. The required point C is then at a distance from A equal to AE. Putting the problem in algebraical terms it amounts to solving the equation

$$x^2 = (1-x) \times 1, \tag{4.1}$$

(AB being taken as unit and AC as x).

Fig. 4.2

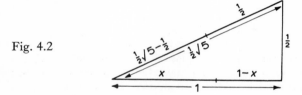

The positive root of this equation is $(\sqrt{5}-1)/2$ which agrees with the geometrical solution.

It should be noticed that AC is the geometric mean between AB and BC. Algebraically this is equivalent to

$$x/(1-x) = 1/x, \text{ another form of equation (4.1).}$$

The similar rectangles formed by 1, x and x, $1-x$ were considered to have particularly pleasing proportions and were therefore widely used in Greek architecture. Hence the names 'golden rectangle, golden mean, golden ratio'.

4.5 A special isosceles triangle

Consider now the isosceles triangle with angles 2α, 2α, and α, where $\alpha = 36°$. Let its sides be 1, 1, and y. Another isosceles triangle with angles α, α, and 3α can be built onto it to make a larger triangle similar to the first but differently oriented.

Fig. 4.3

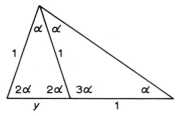

Comparing sides of the similar triangles

$$y/1 = 1/(y+1)$$
$$y^2 = 1 - y. \tag{4.2}$$

Hence y satisfies the same equation as x in section 4.4 and $y/1$ is the golden ratio. Note also that 1 is the geometric mean between y and $y+1$.

There is a useful word 'gnomon' for the additional triangle $(\alpha, \alpha, 3\alpha)$. The term implies that the enlarged figure is the same shape as the original. Clearly the process of adding a gnomon can be repeated indefinitely. The sides of the next gnomon in the series are $y+1$, $y+1, y+2$.

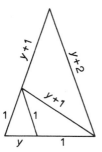

Fig. 4.4

Continuing the sequence, the following lengths appear in order of size as bases of triangles with angles $2\alpha, 2\alpha, \alpha$:

$$y, 1, y+1, y+2, 2y+3, 3y+5, 5y+8, \ldots,$$

and the connection between the golden ratio y and the Fibonacci sequence begins to emerge. Since this sequence is used repeatedly in what follows it is convenient to assign it the symbol S_f and to use the letter f for the golden ratio, both being reminders of the Fibonacci connection.

The sequence S_f could also have been obtained algebraically from the equation (4.2) satisfied by f: $f^2 = 1 - f$. It has already been noted that f, 1, and $f+1$ are in geometrical progression with common ratio

69

$f+1$. Continue the geometric sequence by multiplying repeatedly by $f+1$:

$$(f+1)(f+1)=f^2+2f+1$$
$$=(1-f)+2f+1$$
$$=f+2,$$
$$(f+1)^3=(f+1)(f+2)$$
$$=f^2+3f+2$$
$$=(1-f)+3f+2$$
$$=2f+3.$$

Each term in the sequence reduces to a linear form of f with Fibonacci coefficients.

4.6   Two Fibonacci equations

It has been observed that the equation $x^2+x-1=0$ is a key both to golden section and to Fibonacci numbers. It is time to examine its roots more carefully. They are

$$(-1+\sqrt{5})/2 \text{ and } (-1-\sqrt{5})/2,$$

i.e. $0\cdot 618034\ldots$ and $-1\cdot 618034.\ldots$ Their sum and product are both -1. The companion equation $x^2-x-1=0$ has roots $-\cdot 618034$ and $1\cdot 618034$ to the same degree of accuracy.

In terms of f the four roots are

for $x^2+x-1=0$, f and $-(f+1)$ (4.3)
for $x^2-x-1=0$, $-f$ and $f+1$ (4.4)

Since f satisfies equation (4.3)

$$f(f+1)=1 \tag{4.5}$$

so f and $f+1$ are reciprocals and we can replace $1/(f+1)$ by f and $1/f$ by $f+1$ whenever it suits us to do so.

Another useful form of equation (4.5) is $f^2=1-f$, as we found in the last paragraph. It enables us to reduce any polynomial in f to a linear form. Also since $f=(-1+\sqrt{5})/2$ then $2f+1=\sqrt{5}$, a result used in section 4.11.

4.7   A group relation between the four roots

As a small digression it is perhaps worth noticing that the two equations compound into the quartic equation

$$x^4-3x^2+1=0,$$

and that the four roots of this equation f, $-f$, $f+1$, $-(f+1)$ can be permuted by four functions similar to those discussed in sections 2.12–2.14. The functions are t, $-t$, $1/t$, $-1/t$ which make a Klein's group for substitution.

70

4.8 The Sequence S_f

What happens if we continue the geometrical progression in the reverse direction, using $1/(f+1)$ as a common ratio? We know that $1/(f+1)=f$, so the next term below f in S_f is f^2, i.e. $1-f$.

Continuing:
$$f^3=2f-1$$
$$f^4=2-3f$$
$$f^5=5f-3.$$

The Fibonacci coefficients appear again but mixed up with minus signs. S_f can now be seen as a double-ended sequence, either as powers of f and $f+1$ or as linear expressions in f with Fibonacci coefficients. It is useful to draw up a table in parallel columns.

Table for S_f

Powers	Linear expressions in f
$(f+1)^5$	$5f+8$
$(f+1)^4$	$3f+5$
$(f+1)^3$	$2f+3$
$(f+1)^2$	$f+2$
$f+1$	$f+1$
1	1
f	f
f^2	$-f+1$
f^3	$2f-1$
f^4	$-3f+2$
f^5	$5f-3$

This table suggests that we look again at the bottom end of the Fibonacci sequence and continue it backwards by subtraction of neighbouring pairs.

$$\ldots 5, 3, 2, 1, 1, 0, 1, -1, 2, -3, 5, \ldots$$

In the right hand column of the table for S_f both coefficients of f and the constant terms follow this sequence faithfully.
If we designate its terms

$$\ldots a_3, a_2, a_1, a_0, a_{-1}, a_{-2}, a_{-3}, \ldots (a_0=0),$$

then
$$a_{-n}=(-1)^{n+1}a_n$$

and the general relation in the bottom half of the table for S_f above can be written

$$f^{n+1}=(-1)^n (a_{n+1}f-a_n). \tag{4.6}$$

71

4.9 The convergence of Fibonacci ratios towards the irrational number f

Equation (4.6) of the previous paragraph provides an easy way of proving the convergence of the ratio a_n/a_n+1 towards f, for it may be written in the form

$$a_n - a_{n+1}f = (-1)^{n+1}f^{n+1},$$

i.e. $$a_n/a_{n+1} - f = (-1)^{n+1}f^{n+1}/a_{n+1}.$$

Since $f < 1$ and $a_{n+1} \to \infty$ as n increases, the absolute value of the right hand side can be made as small as we please by taking n sufficiently large. It follows that a_n/a_{n+1} converges towards f as n increases and is alternately greater or less than f according as n is odd or even.

Familiarity with Fibonacci numbers offers excellent opportunities for a display of one-upmanship when motoring on the Continent. It happens that a kilometre is 0·6214 miles, the conversion number being slightly nearer to f (0·6180) than to 5/8 (0·6250) which is the usual approximation. All the Fibonacci ratios are therefore good approximations, so one's fellow-passengers can be stunned by a lightning conversion from 89 kilometres to 55 miles or 11 miles to 18 kilometres.

Exercise

4.1 In section 4.6 it was noted that $f+1$ is a root of the equation $x^2 - x - 1 = 0$. It follows that if we replace $f+1$ by the single symbol F then

$$F^2 - F - 1 = 0,$$

i.e. $$F^2 = F + 1.$$

Use this equation to rewrite S_f as a sequence of linear expressions in F.

4.10 Plotting the lower end of the Fibonacci sequence

The method used in section 4.2 now yields extra coordinate pairs

$$(1,0), (-1,1), (2,-1), (-3,2), \ldots$$

Rather unexpectedly they are found all lying on the same side of the line $y = -fx$ which is at right angles to the line $y = (f+1)x$ which appeared for positive pairs.

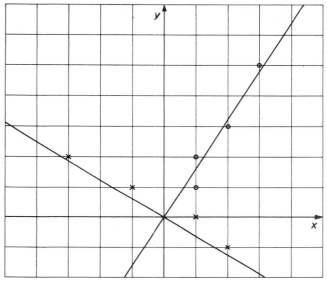

Fig. 4.5

As n increases the points converge on the line from above. The sequence yields no points in the third quadrant and the picture seems incomplete until we remember that this pair of lines belongs to *all* Fibonacci sequences including . . . , -3, -5, -8, . . . With points from this sequence also included, the pattern of points begins to look suspiciously like a pair of rectangular hyperbolas with the two lines as asymptotes. This clue will be followed up later in sections 4.17–19.

To conclude this section it may be said that every point with integral coordinates in the Cartesian plane determines a Fibonacci sequence and that as the ordered pairs of any sequence become larger they define points which converge on one or other of the two lines $y = -fx$ and $y = (f+1)x$. If we choose to reverse the order of the pairs then they will converge on $y = fx$ and $y = -(f+1)x$, another reminder that the roots of the equations $x^2 \pm x - 1 = 0$ are at the heart of the Fibonacci sequences. The two equations combine into the single quartic $x^4 - 3x^2 + 1 = 0$.

4.11   Four interesting power series

The four roots of this quartic can be combined by means of the exponential function to make four interesting series; but first it should be pointed out that S_f is not the only sequence of its kind. Any

73

Fibonacci sequence can be used to generate a system of linear forms in f which are in geometric progression. The other well-known sequence 1, 3, 4, . . ., for example is really a double-ended sequence . . ., -4, 3, -1, 2, 1, 3, 4, . . . from which can be derived the geometric progression . . ., $-4f+3$, $3f-1$, $-f+2$, $2f+1$, . . ., with common ratio $f+1$.

Since $2f+1=\sqrt{5}$ (see section 4.6) each term in this sequence is $\sqrt{5}$ times the corresponding term in S_f. In particular

$$2-f=f\sqrt{5}$$

a result to be used in section 4.17.

Other sequences are not so symmetrical. An initial pair of terms 2, 7 leads to

$$\ldots, -11, 8, -3, 5, 2, 7, 9, 16, \ldots$$

but each has a corresponding geometric progression, with common ratio $f+1$.

Now the exponential series can be brought in:

$$\exp\{-(f+1)x\}=1-(f+1)x+\frac{(f+1)^2x^2}{2!}-\frac{(f+1)^3x^3}{3!}+\cdots$$

$$=1-(f+1)x+\frac{(f+2(x^2}{2!}-\frac{(2f+3)x^3}{3!}+\cdots,$$

$$\exp\{fx\}=\quad 1\;+fx+\frac{(1-f)x^2}{2!}+\frac{(2f-1)x^3}{3!}+\cdots,$$

$$\exp\{fx\}+\exp\{-(f+1)x\}=\quad 2\;-\;x\;+\;\frac{3x^2}{2!}\;-\;\frac{4x^3}{3!}+\cdots.$$

Similarly:
$$\exp\}-fx\}+\exp\{(f+1)x\}=\quad 2\;+\;x\;+\;\frac{3x^2}{2!}\;+\;\frac{4x^3}{3!}+\cdots.$$

The numerators all belong to the sequence . . ., -4, 3, -1, 2, 1, 3, 4, . . . It may also be proved that

$$\exp\{(f+1)x\}-\exp\}-fx=\sqrt{5}\left\{x+\frac{x^2}{2!}+\frac{2x^3}{3!}+\frac{3x^4}{4!}+\cdots\right\},$$

with numerators in the Fibonacci sequence, and this too has a companion equation using the other two roots.

S_f arising from the most familiar of the Fibonacci sequences might itself be called the most perfect sequence of them all since it obeys the formative rule and in addition is a perfect geometric progression yielding plotted points all lying on the line of best fit for the sequences of integers. However it must be remembered that all its members are irrational, whereas the Fibonacci sequence proper has integral members.

74

4.12 The regular pentagon and pentagram

S_f was obtained from a study of a particular isosceles triangle with angles 72°, 72°, and 36° which belongs naturally to the regular pentagon and pentagram.

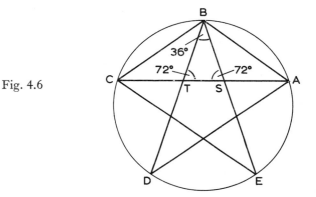

Fig. 4.6

It can be seen at once that T divides CS in golden section, a direct application of 4.5. Also if CT be chosen as unit length

$$TS=f, \quad AS=1, \quad AT=f+1, \quad AC=f+2.$$

Hence S also divides CA in golden section.

4.13 Revolving squares

The golden rectangle has an interesting and well-known property which again gives rise to S_f. Begin with a golden rectangle R, having sides f, 1.

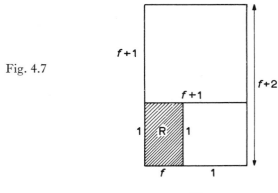

Fig. 4.7

Draw a square on the larger side. The enlarged figure is again a golden rectangle with sides 1 and $f+1$. The square acts as a gnomon. Draw another square on side $f+1$ as shown. A new golden rectangle with sides $f+1$, $f+2$ is formed. Continue the process by adding square gnomons 1, 2, 3, 4, 5 in an anti-clockwise direction round the original rectangle R.

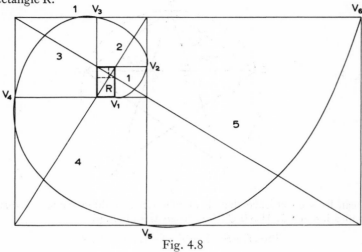

Fig. 4.8

The sides of the various rectangles are always two consecutive terms of S_f. The process can also be continued in a reverse direction. Follow the squares round in a diminishing sequence clockwise and a further square can be cut from R and again from the remainder of R . . . (see Fig. 4.8). These diminishing squares will spiral inwards to a definite limit point at the intersection of the common diagonals of the horizontal and vertical sets of rectangles. Every golden rectangle must contain this point from similarity considerations in the method of construction. Notice that these two common diagonals have gradients $-f$ and $f+1$, and remind us of the two lines of best fit for the Fibonacci sequences.

4.14 The equiangular spiral

In Fig. 4.8, choose a vertex such as V_1 common to a square and a rectangle, and a further point V_2 at the opposite corner of the square. A chain of points V_1, V_2, V_3, . . . may be obtained in this way all of which lie on an equiangular spiral whose centre is the limit point of Fig. 4.8. The spiral is there because the figure contains geometrical progressions not because of its golden section properties. To make this point clear consider a sequence of right-angled triangles rotated repeatedly through 90° anti-clockwise and at each step enlarged in the ratio of the sides containing the right angle.

Fig. 4.9

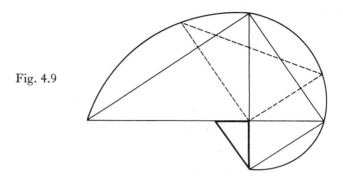

The spiral property is evident even though no golden section is present. This figure could really be regarded as being in a continuous state of development. The triangle has been drawn dotted in an intermediate position and may be thought of as being continually rotated and expanded.

4.15 Rotating rectangles

The rectangle of 4.13 need not be a golden rectangle as any rectangle will illustrate the spiral development. Another well-known case is that of a rectangle whose sides are in the ratio of 1 to $\sqrt{2}$. The next rectangle in the series has sides $\sqrt{2}$ and 2 and the area is doubled at each step. The gnomon is an equal rectangle. An ingenious conjuror's trick is based on this. A playing card swiftly folded in two looks the same shape and has the same picture as a court-card which looks a proper miniature of the old card.

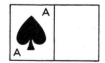 

Fig. 4.10

The trick is repeated to produce a magically diminishing Ace of Spades, simply by folding in half again and again.

77

4.16 Rotation of an equiangular spiral

Although the spiral gets its name from the fact that its tangent makes a constant angle with the radius vector we referred in section 4.14 to another property, namely that successive radii drawn from its pole at regular angular intervals are in geometrical progression.

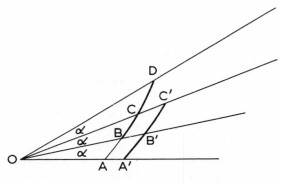

Fig. 4.11

In Fig. 4.11 $\dfrac{OA}{OB} = \dfrac{OB}{OC} = \dfrac{OC}{OD} = k$

and $\hat{AOB} = \hat{BOC} = \hat{COD} = \alpha.$

This definition of a spiral leads on to another interesting property which will be of use in the next chapter. Consider what would happen if the whole plane were transformed by a stretch $1/k$ from the centre O:

> A would go to A′ where OA′=OB,
>
> B would go to B′ where OB′=OC,
>
> C would go to C′ where OC′=OD.

The curve A′B′C′ would be congruent to BCD and the effect of the stretch is to produce a portion of another spiral with centre O, congruent to the original one. The same three points A′, B′, C′, would have appeared as images of B, C, D if the spiral had been rotated clockwise about O through an angle α. Thus to rotate a complete spiral about its pole is to produce the same effect as an expansion from the pole; but it must be noted that A′ is the image of B in the first case and of A in the second. The effect of the two transformations on a limited arc of the spiral is not the same.

4.17 Rectangular hyperbolas associated with the Fibonacci numbers

It was suggested in section 4.10 that points such as (3,5), (5,8) specified by consecutive pairs of Fibonacci numbers might lie on one or other of two rectangular hyperbolas having asymptotes $y=(f+1)x$ and $y=-fx$.

To test this we can rewrite each point as (x,y) referred to the supposed asymptotes as axes. The guess will be verified if the product xy is constant for all points. We shall need to refer to the table for S_f in section 4.8 and the identity $2-f=f\sqrt{5}$ obtained in section 4.11.

The asymptote $y=(f+1)x$ makes an angle θ with the x-axis, where $\theta=\tan^{-1}(f+1)=\cot^{-1} f$. With the usual formulae for change of coordinates resulting from rotation of the axis through an angle θ we can specify the new coordinates of the point (3,5) as follows:

$$x=3\cos\theta+5\sin\theta,$$
$$y=-3\sin\theta+5\cos\theta.$$

Then
$$
\begin{aligned}
xy &=(3\cos\theta+5\sin\theta)(5\cos\theta-3\sin\theta)\\
&=(3\cot\theta+5)(5\cot\theta-3)/\operatorname{cosec}^2\theta\\
&=(3f+5)(5f-3)/(f^2+1)\\
&=(f+1)^4\times f^5/(2-f)\ \text{(see section 4.8, table for } S_f)\\
&=f/f\sqrt{5}\ \text{(see section 4.11)}\\
&=1/\sqrt{5}.
\end{aligned}
$$

In the same way for the point (5,8)

$$
\begin{aligned}
xy &=(5f+8)(8f-5)/(2\text{-}f)\\
&=-1/\sqrt{5}.
\end{aligned}
$$

It is clear that successive coordinate pairs indicate points alternately on the two hyperbolas $xy=\pm 1/\sqrt{5}$.

Now the combined equation of the two asymptotes referred to the original axes is

$$(y-(f+1)x)(y+fx)=0,$$

i.e.
$$y^2-xy-x^2 \qquad =0, \qquad (4.7)$$
i.e.
$$(y/x)^2-(y/x)-1 \qquad =0.$$

If we now put $y/x=t$ the equation comes back to the characteristic Fibonacci form

$$t^2-t-1=0.$$

Also by the standard property of a general conic, equation (4.7) for the asymptotes implies an equation for each hyperbola of the form

$$y^2-xy-x^2=k. \qquad (4.8)$$

79

To find k and verify that the Fibonacci point (u_n, u_{n-1}) does indeed satisfy equation (4.8) we can test for special points following the usual notation for the sequence, u_n being the nth term:

$$u_0=0,\ u_1=1,\ u_2=1,\ u_3=2,\ u_4=3,\ \ldots$$

For $n=1$, $1^2-1\cdot1-1^2=-1$;
$\ \ \ \ \ n=2$, $2^2-2\cdot1-1^2=\ \ \ 1$;
$\ \ \ \ \ n=3$, $3^2-3\cdot2-2^2=-1$;
$\ \ \ \ \ n=4$, $5^2-5\cdot3-3^2=\ \ \ 1$.

The relationship when stated generally gives

$$(u_{n+1})^2-u_{n+1}u_n-u_n{}^2 =(-1)^n,$$
i.e. $(u_{n+1})^2-u_n(u_{n+1}+u_n)=(-1)^n,$
i.e. $(u_{n+1})^2-u_nu_{n+2}\ \ \ \ \ \ \ =(-1)^n.$

By this route we have arrived at one of the well-known properties of the Fibonacci sequence (see Vorob'ev, *Fibonacci Numbers*).

4.18 Other Fibonacci sequences and hyperbolas

The method used in the last section can also be adapted for any other sequence. Consider for example the less symmetrical sequence A generated by the numbers 2,5.

$$A=\ldots,\ -1,3,\underline{2},5,7,12,\ldots$$

The product of coordinates of a typical point $(5,7)$ referred to the asymptotes as axes would be

$$(5f+7)\ (7f-5)/(2-f).$$

Because of the lack of symmetry between the two ends of the sequence A it becomes necessary to consider also the allied sequence B formed from $-5,2$.

$$B=\ldots,\ -12,7,\underline{-5,2},-3,-1,\ldots$$

From A and B we can derive two geometrical progressions and write down corresponding terms with similar coefficients:

A	B
$-f+3$	$-3f-1$
$3f+2$	$2f-3$
$2f+5$	$-5f+2$
$5f+7$	$7f-5$
$7f+12$	$-12f+7$

The sequence under A is an increasing one with common ratio $f+1$; that under B consists entirely of negative terms, decreasing in absolute size with common ratio f. Since f and $f+1$ are reciprocals the absolute value of the product of corresponding terms is constant. Hence the point (5,7) and all other points determined by two consecutive terms of A lie on one of two hyperbolas as before. It can quickly be verified by taking sample points that the equation of the two hyperbolas referred to the original axes is

$$y^2 - xy - x^2 = \pm 11$$

For example, if the point is (7,12), then

$$12^2 - 7 \cdot 12 - 7^2 = 144 - 84 - 49$$
$$= 11.$$

Again the method can be generalised for a sequence formed from any two numbers a, b by considering the two sequences A,B:

$$A = \quad a, b, \; a+b, \; a+2b, \; 2a+3b, \; \ldots,$$
$$B = -b, a, \; a-b, \; 2a-b, \; 3a-2b, \; \ldots.$$

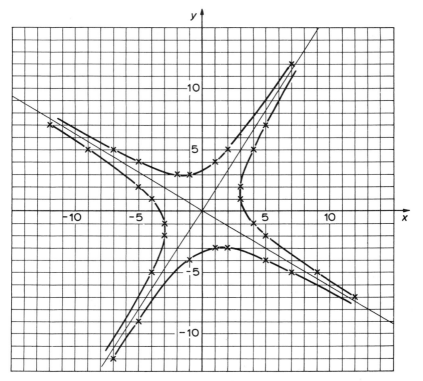

Fig. 4.12

81

In this case all points specified by two consecutive elements of either A or B lie on one or other of the hyperbolas

$$y^2 - xy - x^2 = \pm(b^2 - ab - a^2).$$

Still further, every such pair of hyperbolas contains sets of points derived from four allied Fibonacci sequences. For example, the hyperbolas $y^2 - xy - x^2 = \pm 11$ (Fig. 4.12) contain points derived from the four sequences:

$$
\begin{array}{rrrrrrrrrrrr}
.., & 9, & -5, & 4, & -1, & 3, & 2, & 5, & 7, & 12, \ldots \\
.., & & & -12, & 7, & -5, & 2, & -3, & -1, & -4, & -5, & -9, & -14, \ldots \\
.., & & & 12, & -7, & 5, & -2, & 3, & 1, & 4, & 5, & 9, & 14, \ldots \\
.., & -9, & 5, & -4, & 1, & -3, & -2, & -5, & -7, & -12, \ldots \\
\end{array}
$$

4.19 Extension of the term Fibonacci sequence

So far we have assumed that the sequences are derived from integers. There is nothing in the last section which would not be equally true for any pair of real numbers. It would then follow that every point (a,b) in the Cartesian plane determines a sequence and belongs to one of a pair of hyperbolas such that

$$x^2 - xy - y^2 = \pm(a^2 - ab - b^2).$$

4.20 A matrix treatment of Fibonacci numbers

For those familiar with matrix algebra there is a different treatment of Fibonacci numbers which is relevant to some sections of this chapter.

First let us again set out the terms of the sequence using a standard notation:

$$
\begin{array}{ll}
u_0 = 0, & u_4 = 3, \\
u_1 = 1, & u_5 = 5, \\
u_2 = 1, & u_6 = 8, \\
u_3 = 2, & u_7 = 13.
\end{array}
$$

Then as in section 4.2, let us designate a sequence of points in the Cartesian plane by pairs of consecutive numbers

$$P_0 = (0,1),\ P_1 = (1,1),\ P_2 = (1,2),\ \ldots,$$

but let us change the notation slightly so that P_0, P_1, P_2 have position vectors $\begin{pmatrix} 0 \\ 1 \end{pmatrix}, \begin{pmatrix} 1 \\ 1 \end{pmatrix}, \begin{pmatrix} 1 \\ 2 \end{pmatrix}$.

This could be summed up as a one-one mapping

$$P_n \longleftrightarrow \begin{pmatrix} u_n \\ u_{n-1} \end{pmatrix}.$$

Next, if $\begin{pmatrix} a \\ b \end{pmatrix}$ stands for any vector in the series we look for a matrix M which will transform it into the next member of the series $\begin{pmatrix} b \\ a+b \end{pmatrix}$. It is clear that $\begin{pmatrix} 0 & 1 \\ 1 & 1 \end{pmatrix}$ will have this effect and that no other matrix will do so. Since M is independent of a and b, it will transform any point in the series $P_0, P_1, \ldots P_n$ to its next-door neighbour. Its effect on the series is the mapping

$$P_n \longrightarrow P_{n+1}.$$

It follows that if the transformation M is used twice over it will produce the mapping

$$P_n \longrightarrow P_{n+2}.$$

Confirming this in a particular instance

$$M^2 = \begin{pmatrix} 0 & 1 \\ 1 & 1 \end{pmatrix} \begin{pmatrix} 0 & 1 \\ 1 & 1 \end{pmatrix}$$

$$= \begin{pmatrix} 1 & 1 \\ 1 & 2 \end{pmatrix},$$

and

$$\begin{pmatrix} 1 & 1 \\ 1 & 2 \end{pmatrix} \begin{pmatrix} 3 \\ 5 \end{pmatrix} = \begin{pmatrix} 8 \\ 13 \end{pmatrix},$$

i.e.

$$M^2 \begin{pmatrix} u_4 \\ u_5 \end{pmatrix} = \begin{pmatrix} u_6 \\ u_7 \end{pmatrix}$$

so that M^2 sends the point P_4 to P_6. The power series M, M^2, ... can be continued indefinitely:

$$M^3 = M \times M^2$$

$$= \begin{pmatrix} 0 & 1 \\ 1 & 1 \end{pmatrix} \begin{pmatrix} 1 & 1 \\ 1 & 2 \end{pmatrix}$$

$$= \begin{pmatrix} 1 & 2 \\ 2 & 3 \end{pmatrix}.$$

We need not go very far with the M series to discover that it inevitably generates Fibonacci numbers in a steady sequence. One advantage of this is that by repeated squaring we can leap up the sequence in longer and longer bounds, so reaching high Fibonacci numbers very quickly. Only six stages would give us the 63rd, 64th, and 65th Fibonacci numbers.

All these results can be combined in a table which contains the sequence of matrices as overlapping elements:

n	0	1	2	3	4	5	6	7...
u_n	0	1	1	2	3	5	8	13...
u_{n+1}	1	1	2	3	5	8	13	21...

$$\text{M} \quad \text{M}^2 \quad \text{M}^3 \quad \text{M}^4 \quad \text{M}^5 \quad \text{M}^6 \quad \text{M}^7 \quad \ldots$$

The matrix M^3 is indicated in this table as $\begin{pmatrix} 1 & 2 \\ 2 & 3 \end{pmatrix}$, a result we determined previously. It is reasonable to state, though we have not rigorously proved it, that

$$\text{M}_n = \begin{pmatrix} u_{n-1} & u_n' \\ u_n & u_{n+1} \end{pmatrix}.$$

Consider now the triangles $\text{O P}_0 \text{P}_1$ and $\text{O P}_1 \text{P}_2$.

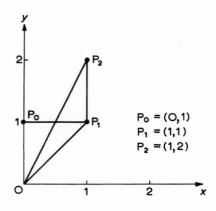

$$\text{P}_0 = (0,1)$$
$$\text{P}_1 = (1,1)$$
$$\text{P}_2 = (1,2)$$

Fig. 4.13

Both have area $\frac{1}{2}$ but if we apply the well-known determinant formula for the area of a triangle

$$\text{area} = \tfrac{1}{2} \begin{vmatrix} x_1 & y_1 & 1 \\ x_2 & y_2 & 1 \\ x_3 & y_3 & 1 \end{vmatrix}$$

84

for $O P_0 P_1$, we get

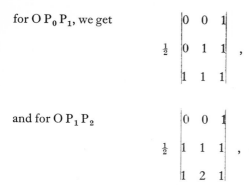

$$\tfrac{1}{2} \begin{vmatrix} 0 & 0 & 1 \\ 0 & 1 & 1 \\ 1 & 1 & 1 \end{vmatrix} \; ,$$

and for $O P_1 P_2$

$$\tfrac{1}{2} \begin{vmatrix} 0 & 0 & 1 \\ 1 & 1 & 1 \\ 1 & 2 & 1 \end{vmatrix} \; ,$$

leading to $-\tfrac{1}{2}$ and $\tfrac{1}{2}$ respectively.

It is convenient to take $O P_1 P_2$ with positive area as the initial triangle and then to consider the series of triangles $O P_1 P_2$, $O P_2 P_3$, ... Since multiplication by M leaves the zero vector, corresponding to 0, unchanged it transforms the triangle $O P_1 P_2$ into $O P_2 P_3$ and so on. The determinant of M is -1 and this tells us two important facts: (i) that the area of the triangle is unchanged and (ii) that the sense of description of the triangle O to P_1, to P_2 is reversed from anti-clockwise to clockwise.

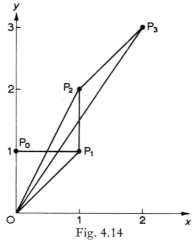

Fig. 4.14

After, say, five steps the name of the triangle would be OP_6P_7 and its area would be $-\tfrac{1}{2}$.

4.2   (a) Verify this last statement by finding the area from determinant (1) and write down a relation between $u_5\, u_7$ and u_8.
   (b) Obtain a similar relation between $u_6\, u_7$ and u_8.
   (c) What is the general relation connecting u_n, u_{n+1} and u_{n+2}?
   (d) Compare your result under (c) with that at the end of section 4.17.

The triangle OP_nP_{n+1} having constant area gets more and more elongated as $n \longrightarrow \infty$. Regarding OP_n as base, its corresponding altitude must $\longrightarrow O$. The angle P_nOP_{n+1} is also shrinking towards zero, a fact that we observed empirically in section 4.2. Looking back over previous results we know that P_n lies alternately on either side of the line $y=(f+1)x$ and approaches it asymptotically. It is relevant to enquire about the effect of M on a vector OP lying along this line. Such a vector would be of the form

$$\begin{pmatrix} a \\ a(f+1) \end{pmatrix}.$$

$$\begin{pmatrix} 0 & 1 \\ 1 & 1 \end{pmatrix}\begin{pmatrix} a \\ a(f+1) \end{pmatrix}=\begin{pmatrix} a\,(f+1) \\ a\,(f+2) \end{pmatrix}$$
$$=\begin{pmatrix} a\,(f+1) \\ (a\,)f+1)^2 \end{pmatrix}$$
$$=(f+1)\begin{pmatrix} a \\ a(f+1) \end{pmatrix}.$$

We see that OP is multiplied by a scalar factor $(f+1)$, and this means that OP is an eigenvector of M. The same fact would emerge if we obtained the *characteristic equation* of the matrix M and it is not surprising that this turns out to be the familiar equation underlying the Fibonacci sequence

$$x^2+x-1=0,$$

of which the roots are $f+1$ and $-f$, so that

$$\begin{pmatrix} a \\ a(f+1) \end{pmatrix} \text{ and } \begin{pmatrix} a \\ -af \end{pmatrix}$$

are both eigenvectors of M.

We are now in a position to say exactly what transformation M represents. It is a stretch of amount $f+1$ along the line $y=(f+1)x$ combined with the reciprocal stretch (or contraction) f along the line $y=-fx$ and a reflection in the line $y=(f+1)x$.

Exercises

4.3 Verify by application of the matrix M that $\begin{pmatrix} a \\ -af \end{pmatrix}$ is an eigenvector of M.

4.4 Examine the effect of the matrices $\begin{pmatrix} 0 & -1 \\ -1 & -1 \end{pmatrix}$ and $\begin{pmatrix} -1 & 1 \\ 1 & 0 \end{pmatrix}$ on various Fibonacci vectors such as $\begin{pmatrix} 2 \\ 3 \end{pmatrix}$.

What transformations do these matrices represent? Interpret your results with the aid of a diagram of two conjugate hyperbolas similar to Fig. 4.12.

5

BIOLOGICAL ILLUSTRATIONS OF FIBONACCI SEQUENCES

5.1 Fibonacci numbers in nature

Many interesting accounts have been written of the ways in which Fibonacci numbers keep cropping up in biological situations. We began the last chapter with the development of a rabbit colony, admittedly rather an artificial example. The ancestry of a male bee has been described in Professor F.W. Land, *The Language of Mathematics*. Here is the drone's family tree, taking into account that he comes from an unfertilised egg of a queen bee:

Fig. 5.1

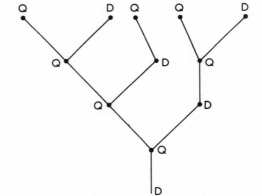

Successive generations of ancestors follow the Fibonacci sequence; this assumes, however, that all queens in the ancestry are distinct, which is not necessarily true.

5.2 Growth of a committee

Parkinson's Law relates to the growth of staff in a bureaucracy. In a neighbouring chapter the author also describes the development and expansion of committees. Here is a constitution for the planning committee of a big organisation as Fibonacci might have devised it.

Rule 1. The first member shall retire or become an advisory member after one year. Every other member shall serve as a voting member for exactly two years and shall then retire or else continue to serve but as an advisory member.

Rule 2. In each year of service each voting member shall nominate one new member to the committee.

Rule 3. Each member shall have a rank number according to seniority of appointment. Rank 1 will be given to the first member and 2, 3, 4 . . . to other members as they are appointed. Members shall retain this rank after retirement.

Rule 4. Voting members shall appoint new members in strict order of seniority.

Rule 5. In years of even date new members shall take their seats at the committee table on the left of the member appointing them. In odd years they shall be seated on the right of their sponsors.

Rule 6. An advisory member shall continue to attend meetings of the planning committee: if unable to do so owing to pressure of work on other committees to which he may belong, he shall automatically be appointed to the rank of retired member.

The following diagrams show how the committee would look at the beginning of successive years I, II, III, . . .

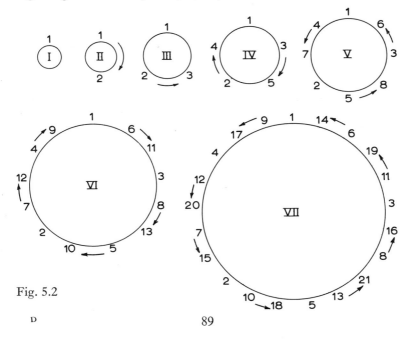

Fig. 5.2

There are several points of interest to note in the development

(i) The total number of members at the beginning of each year follows the Fibonacci sequence.

(ii) It can be seen from the diagram for year VI that there are 5 new members, 5 nominating members, and 3 advisory members. In year VII the pattern will be repeated with 8, 8, 5.

(iii) The member of rank 2 is at roughly the same angular distance from 1 throughout the development. His station is at an angular distance from 1 of $2\pi \times p$ where p follows the converging sequence $1/2, 1/3, 2/5, 3/8$. . . . Measured in degrees instead of radians, the angle is $360° \times p$.

(iv) Other members in order of seniority are found at the points of a star polygon.

Fig. 5.3

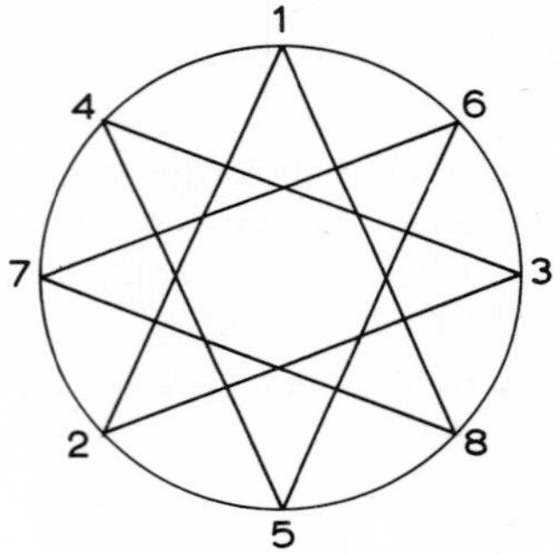

5.3 Leaf arrangement in plants

This rather elaborate constitution for an imaginary committee has been described in such detail because it might perhaps provide a clue to the fascinating and perplexing problem of phyllotaxis or leaf arrangement in plants. The main facts are well-known:

(i) In a great many plants leaves appear in a spiral pattern round the stem. Biologically this is a good arrangement because if they were arranged above each other the lower leaves would be screened from sun and light.

(ii) In many cases the angular distance between successive leaves is determined by one of the ratios $1/2, 1/3, 2/5, 3/8$, . . . of a revolution.

(iii) The numbers of petals in very many flowers which have the characteristic described in (ii) are Fibonacci numbers.

(iv) In certain flower heads which develop radially along a flat surface as in a sunflower or pyrethrum, the florets tend to fall into two spiral patterns, curving in opposite directions. The spirals may be in families of 5 crossing 8, or 8 crossing 13, and sunflowers have been found with much larger Fibonacci pairs such as 55 crossing 89.

(v) The scales of a pine cone are arranged in spiral patterns round the longitudinal axis. There are examples of 3 to 5, 5 to 8, and 8 to 13, each tree having its own characteristic combination.

5.4 Biological background for a hypothesis

A tentative suggestion for an explanation of the widespread occurrence of Fibonacci numbers in nature is put forward in later sections of this chapter. The present section is concerned with the necessary biological background of the discussion and in it I have drawn heavily on the writings of eminent biologists. The hypothesis was formulated after reading an account of phyllotaxis by F.J. Richards, and this is an attempt to put together the main sources of ideas that have come into my mind since then, during a course of reading with a view to substantiating the hypothesis.

(a) *Cell division*

Plants grow by division and also by enlargement and maturation of cells. There are two kinds of cell division, mitosis (asexual) and meiosis (sexual). We are concerned here only with vegetative growth which comes from mitotic division. Details of the process are irrelevant here but its general effect is described in P.B. Weisz and M.S. Fuller, *The Science of Botany*, p.372–8.

> 'In Protista and Metaphyta, cell division consists of at least two separate processes: one is division (actually duplication) of the nucleus, and the other is cleavage of the cytoplasm into two parts. The process of nuclear division is known as mitosis, that of cytoplasmic division as cytokinesis. . . . Cytokinesis terminates roughly when mitosis terminates. The net result of both processes is the division of the cell into two cells containing *precisely* identical gene sets incorporated in identical chromosome sets and *approximately* equal quantities of all other cellular constituents. Consequently the structural and functional potential of both daughter cells is the same as that of the mother cell.'
>
> 'The rates of cellular reproduction vary greatly. Among multi-cellular plants, the most intense rates of cell division occur in embryonic stages, the least intense in old age. Apical and other permanently embryonic cells retain a fairly rapid rate of division throughout the life of a plant. By contrast, differentiated cells divide only rarely in the adult. In general, the more highly specialised a cell, the less frequently it divides, and vice versa.'

Rates of cellular reproduction are also mentioned in Strasburger's, *Textbook of Botany*, p.314.

'Cell divisions often take place with a definite frequency, the period sometimes being diurnal, but also occasionally such that several divisions occur within twenty-four hours.'

(c) *Meristem*

The part of a plant responsible for forming its basic structure is the *meristem*, situated mainly at the apex or growing-point of the main stem, side-shoots, or roots-tips. The word is of Greek origin and means divisible.

'In active meristems a part of the products of cell-division remain meristematic, the *initiating cells*, and others develop into the various tissue elements, the *derivatives* of the initiating cells.' (K. Esau, *Plant Anatomy*, p.69).

'The tissues of a plant develop and mature gradually, in pace with the maturation of the various cell types. Taking the embryo of a flowering plant as an example, we find that, at an early developmental stage, such an embryo consists of relatively unspecialised cells which are all more or less alike. The cells at the upper and lower tip of the embryo remain in this unspecialised condition indefinitely, even while the rest of the embryo develops into an adult plant. These permanently embryonic cell groups are called apical meristems, and we may distinguish between an apical shoot meristem at the upper tip and an apical root meristem at the lower tip.' (Weisz and Fuller, p.80).

General patterns of meristematic growth are described in E.J.H. Corner, *The Life of Plants*, p.44.

'Plant cells are strict in the direction in which they divide. The partitions between daughter-cells are not laid down haphazardly. They follow the direction in which the nucleus divides and the partition wall is placed between the daughter-nuclei at right angles to their line of separation. Generally, too, as shown by D'Arcy Thompson, the partitions occupy a minimum area between the divided cells. A cell may lengthen away from the rock or it may stretch sideways over the rock; some seaweeds have one manner of growth, others the other, while the massive fucoids and laminarians combine both. If the cell grows most actively on the free side away from the attached base – that is if the cell grows apically – the new partition separates a basal attachment cell from an apical cell seated upon it and if the process is repeated a row of cells projecting into the water will be formed. If alternatively a cell grows sideways in one direction and lengthens over the rocks a decumbent row will be formed. If then, these cells spread sideways, other partitions will form to make a plate of cells. Similarly a plate can arise through apical growth accompanied by lateral growth from the cells of the filament and a membraneous frond swaying in the water will arise.'

Another description of cellular proliferation is given in A. E. Needhams *The Growth Process in Animals*, pp.42, 44.

'The number of possible spatial patterns based on localised growth zones is strictly limited if we accept the mechanical sanction that new cells can be placed only where there is space for them, usually at surfaces. Growth zones then must be at the most two-dimensional or laminar and they may be one-dimensional in the form of a line or strip or of a closed ring. A ring-shaped zone can grow simply as a ring in the circumferential direction only or it can grow both in this and the radial direction forming a disc or plate. When a growth ring grows circumferentially as well as perpendicularly to its plane then conical structures are formed.'

(c) *Apical growth*

Corner (p.55) also gives a graphic description of apical growth.

'Simple seaweeds are filamentous. Some consist of a single row of cells up to one or a few inches long and all the cells except the basal hold-fast can divide and add to the length of the filament. Others are branching filaments. . . . But the branching filament has established a manner of growth which persists through most forms of higher plants. It is apical growth. Instead of all the cells dividing by transverse walls and young cells with thinner walls making weak links in the chain, cell-division is restricted to the terminal or apical cell of the filament. Branching then arises by outgrowths from the other cells to establish lateral filaments with their own apical cells. The whole structure grows apically at its tips, branches sub-apically, and progressively strengthens itself with thicker walls and hyphae downwards. The apical cell somehow commands the daughter-cells that it has formed and prevents them from dividing, in spite of their growth, but it allows them at certain distances below it to branch out and form new filaments. . . . In land plants the connection between apical growth and apical dominance is well known and is attributed to certain chemical substances that spread from the apex down the plant to inhibit branching.'

93

(d) *Initials*

Some plants have one easily recognisable initial cell at the top of the apex called the *apical initial* and this where it exists is said to give rise to the whole of the stem, leaves, shoots, etc. and to play a vital role in the orderly development of the plant. Other more complicated plants rely on a group of initials rather than one special one and these initials are less easy to identify with certainty.

K. Esau (p.90) writes:

'An initial or initiating cell is a cell that divides into two sister cells, one of which remains in the meristem and the other is added to the meristematic tissues that eventually differentiate into the various tissues characteristic of the plant. The cell remaining in the apical meristem functions as an initial like its precursor. Investigators visualise the involvement of polarity, and a consequent cytologic differentiation in the divisions into an initial and its derivative; at the same time they agree that the status of a cell as an initial depends on its position in the protomeristem and that the initial may be displaced by another cell and then differentiate into a body cell.'

F.E. Fritsch, *The Structure and Reproduction of the Algae*, p.351, refers to initials as follows:

'Dichotomus branching generally takes place in *Fucus* soon after flattening has become apparent. It is initiated by a segment (usually representing half the initial cell) from one of the lateral faces assuming the form and function of an apical cell. According to Oltmanns the two apicals thus constituted cut off a considerable number of large segments . . . which for a time undergo little division so that the initials are at first difficult to distinguish, although later they stand out clearly. Each apical cell then proceeds to form a limb of the dichotomy, while the intervening segments produce a mound of tissue which divides the apical slit into two.'

The next passage comes from F.A.L. Clowes, *Apical Meristems*, p.18.

'Obviously in shoots without a single apical cell there are no cells which are totipotent in the sense of giving rise to all the tissues of the shoot. It is, however, likely that some of the cells in the apex of *Selaginella willdenovii* and similar plants are *totipotent* in the sense of being capable of giving rise to any tissue of the shoot. Such plants may be thought of as possessing a small group of cells at the apex which replace the single apical cell in that all the future cells of the shoot apex are derived from them. Such a

94

group of cells is called a promeristem in this book, and the constituent cells are called initials. . . . It is not supposed that any cell is an initial because of any inherent property not possessed by other cells of the meristem. The status of an initial is given by the position of the cell in the meristem. Nor is it supposed that an initial is necessarily indispensable to the future growth of the shoot. Another initial or some other cell takes its place if it is injured.'

(e) *Differentiation*

'As a result of the division and growth of the embryonic cells at the growing point, and the subsequent extension and differentiation of certain zones, specific organs begin to be developed. This is only possible because the continued development of the innumerable individual cells conforms to a general overriding plan. . . The origin and functioning of the formative and directing forces, the existence of which is already evident in the embryonic growing point, is completely unknown. It is very significant that the embryonic cells initially possess the capacity to give rise to all the parts and tissues of the differentiated plant body. They are therefore, like the zygote, totipotent. After a certain stage of development, however, individual cells or groups of cells become determined. In consequence only a few of the potentialities originally present in the cell are deployed and developed, and the division of labour between cells begins.' (Strasburger, p.318).

(f) *Primordia of leaves and buds*

N.J. Berrill, *Growth, development and pattern*, pp.420, 444, describes the inception of primordia.

'Primordia arise on the apex in a definite and precise pattern, i.e. with phyllotaxis.'
'A primordium of any kind, vegetative or floral, is an approximately circular area or disc of tissue which forms in the first or second subdermal layer of the apical meristem and exhibits a relatively high rate of growth'.

Buds are described in C.W. Wardlaw, *Organisation and Evolution of Plants*, p.251, as follows:

'A growing lateral bud is an embryonic shoot or a shoot in miniature, i.e. it consists of a short polarised vascularised axis with a distal meristem which gives rise to lateral organs. Buds are potentially capable of unlimited growth and develop a tissue pattern similar to that of the main axis. The buds in many species

95

are radially symmetrical but even in species with a dorsiventral axis, the extreme dorsiventrality characteristic of leaves is relatively rare in buds'.

(g) *Plastochrone*

'Whatever the manner of formation, a primordium represents a local region of increased growth rate relative to the rate in its immediate surroundings. Whatever the sequential pattern, the process of initiation is rhythmic, and the time interval between the *initiation* of one leaf primordium and that of the next is known as a plastochrone'. (Berrill, p.417).

(h) *Zones*

Wardlaw (p.230), describing apical organisation and the inception of pattern, writes as follows:

'The shoot apex, which consists of the apical meristem and the subapical and maturing regions, is a harmoniously developing whole or continuum. In the apical meristem itself, one may recognise that there is substantial specific organisation and physiological differentiation: it probably comprises several distinct though inter-related and integrated regions or zones, namely, (i) the *distal*, (ii) the *sub-distal* and (iii) the *organogenic*, zones. . . . While it is in the third of these zones that leaf primordia can first be observed as localised outgrowths, their actual inception as growth centres must take place somewhat higher up in the apex, i.e. in zone (ii). The sub-distal zone is accordingly a region of great importance.'

Esau gives a somewhat similar description but combines zones (ii) and (iii) above into the *peripheral* zone in which the leaf primordia arise; however she also mentions that the border between the distal zone and the peripheral zone is one of intense meristematic activity. She continues the description (p.97):

'The next development in the interpretation of apical meristem resulted from the efforts of Buvat and his students to obtain a unified concept of growth of the meristem. Counts of mitoses and cytological, histochemical and ultrastructural studies served to formulate the theory that the distal zone is relatively inert during vegetative growth and that the real initial zone is the peripheral one where leaf primordia arise. The distal zone received the appellation of waiting meristem (meristem d'attente) because it was said to be waiting for the change from vegetative to reproductive activity. The peripheral zone became the *initiating ring* (anneau initial). . . . The concept was later somewhat modified in that variations in degree of inactivity of the distal zone in relation to the size of the apex and its stage of development came to be recognised.'

(j) *Growth centres and gradients*

There seems to be much controversy as to whether a cell is destined from its inception to become the initial of a primordium or acquires that status later because it happens to occupy a particular position in the plant body at a particular time when a new phase of development is due to begin. In more technical terms its role as a growth centre might depend on its position in a 'field' determined by a number of 'gradients'.

'A question that is left unanswered in gradient theories of organisation is how centres of active metabolism or growth centres originate. The field concept which has been elaborated in great detail by zoologists has been adopted to only a minor extent by botanists.

'Schoute (1913) and Richards (1948, 1951) have applied what is virtually the field concept to the problem of leaf-determination and phyllotaxis in plants.

'In experiments the writer (Wardlaw, 1949, 1952b) showed that, with modifications, Schoute's idea of growth centres affords a basis for a unifying and comprehensive conception of morphogenetic processes at the shoot apex and for the regulated and harmonious formation of the leafy-shoot. The problem then becomes one of envisaging some coherent system which is capable of giving rise to an orderly sequence of growth centres.' (Wardlaw, p.29–33).

(k) *Universality of some components of morphogenesis*

Wardlaw quotes E.W. Sinnott as saying that several morphogenetic components are of very wide occurrence in the Plant Kingdom, mentioning particularly polarity, differential gradients, symmetry and spirality. Both Needham and Berrill remark on the fact that phyllotactic patterns are common to both plant and animal forms of life.

'Another type of spatiotemporal pattern is the phyllotactic, so called because it is seen most characteristically in the order of formation of new leaves on a shoot. The new items are discrete, and therefore growth is spatially discontinuous, but they do appear in regular succession, following a spiral path along the stem of a plant or the body of an animal. In *Hydra* new buds appear along the spiral with a radial angle of 120° between successive members. It is therefore said to have a phyllotactic ratio of 1/3, three buds in

97

each turn around the body, the n-th bud always being vertically above the $(n-3)$th. The possible phyllotactic ratios are members of a Fibonacci series: $1/2$, $1/3$, $2/5$, $3/8$ . . . both numerator and denominator of each term being the sum of those of the previous two terms.' (Needham, p.63).

The reason for quoting this passage is that it refers not to plants but to a colonial organism *Hydra*. Like some others of its type it looks much more like a plant than an animal with a central stalk, branches and buds which are eventually released as new individuals. Berrill describes the *Hydra* and allied organisms in detail and emphasises its resemblance to plant forms.

'The pattern of development of a multicellular organism is generally one of three types: (1) primarily radial; (2) predominantly bilaterally symmetrical; or (3) essentially repetitive, as in segmented animals, colonial organisms and herbaceous plants.' (Berrill, p.158).

'The phenomena of morphogenesis and polymorphism in the development of the shoot and floral apex of vascular plants clearly have much in common with, for example, the successive formation of manubrial buds in a medusa and the processes of rhythmic growth as exhibited in segmentation and strobilation. There are also significant parallels between the developmental polymorphism of hydranth and medusa on the one hand and of vegetative and floral development on the other.' (Berrill, p.413).

(l) *Polarity*

'We can envisage that in each cell certain substances or structures are distributed unequally in a polar manner, so that a concentration of gradient comes into existence. Each dividing wall formed transversely to this gradient inevitably divides the cell into two daughter-cells of unequal potentiality. The cell cut off towards the apex retains the greater part of the meristematic activity, while that towards the base is capable of undergoing only a limited number of divisions. Frequently the basal cell is capable of dividing only after a short or a prolonged resting period, or after a regeneration phase. This is an example of the phenomenon of unequal division.' (Strasburger, p.61).

(m) *Ageing of cells*

'It appears that any living cell must inevitably change with time and after a definite life span, different in different individuals, must die'. (Strasburger, p.327).

'Regardless of where, when, or how we examine the structure

of any cell, we ultimately find it to consist enitrely of chemicals; atoms and molecules. And regardless of what particular function of a cell we examine, that function is ultimately always based on the properties of the cellular atoms and molecules.

'But note that the atomic and molecular composition of a living cell is never static or fixed. As we shall see, new materials enter a cell continuously, wastes and manufactured products leave continuously, and substances in the cell interior are continuously transformed chemically and re-distributed physically. Consequently, as cells *function*, such functioning invariably means *change* of structure and composition. And we must keep in mind that living composition and structure, no less than living function, have important progressive, historical aspects which cannot be divorced from the dimension of time.' (Weisz and Fuller, p.39).

5.5 A hypothetical explanation

Among all the main factors of morphogenesis it is clear that time is of tremendous importance. There are many references to the continuous development of cells: like multicellular organisms they mature, grow old, and die. But in speaking of a cell, what is its date of birth? At what point in time does it individual existence begin? I would like to suggest that a cell retains its identity in some sense even when mitotic division takes place. Instead of the result of such a division being two daughter-cells, might not the relation be rather that of mother and daughter, the mother being the original cell, now one generation older, and the other a new cell coming into existence for the first time? It has been established beyond doubt that some at least of the constituents of a cell divide by replication not by fission, and the replica must be younger than the original.

If this hypothesis is viable its corollary might be a reassessment of the function of the single apical cell in any plant which possesses one. Instead of the apical cell remaining permanently at the apex, cutting off successive segments to form derivatives which move further and further away from the apex as later segments intervene, it might be the case that the mother cell is the one which moves off, leaving the new-born daughter in the apical position. This would account for the puzzling fact that the apical cell seems to have a perpetual youthfulness, remaining always in an embryonic state. In the case of the simple filamentous seaweed described by Corner there would then be a straight succession of cells of increasing age down the filament to the holdfast, with the youngest and mechanically weakest at the top. This particular form of apical growth assumes that the cells in the filament divide only once and always in the same direction, but the possession of an age-label would allow many variations, even in a simple filament, if cells divided more than once and not always in the same direction. If each cell divided repeatedly after missing one generation at the beginning like the Fibonacci rabbits and if the generations alternated in direction the sequence would be:

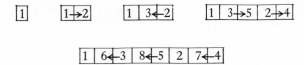

and this would produce an effect very like what is often described as diffuse growth along a filament of cells.

A very similar effect would result from using the rule suggested for the Fibonacci committee: each cell dividing exactly twice in alternate directions (except for the first which is to divide once only) before going on to a further stage of maturity.

Curiously enough if either of these two principles of development operates in such a way that the sequence of cells forms a *ring* it leads to exactly the same cyclic arrangement, the one already depicted in Fig. 5.3.

Another interesting habit of growth which might possibly be a variation on this theme is described by both Strasburger and Fritsch. In the paragraph quoted in subsection (k) under 'Polarity', Strasburger continues:

'In the brown algae belonging to the Sphacelariaceae, which have uninucleate cells, the polarised differentiation of the protoplasm in the large apical cell is made evident by the difference in colour between the apical and the basal regions. . . . Unequal division results in new cells being added basipetally. These cells can divide further, but of prime importance is the first transverse division, since this produces an upper nodal cell and a lower internodal cell. Further division does not take place until after a prolonged resting period, but eventually the nodal cell undergoes longitudinal divisions and gives rise to branches. While this is occurring the internodal cell contributes to the accompanying extension growth.'

100

The section is illustrated diagrammatically:

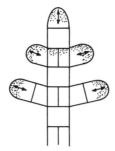

Fig. 5.4

The whole process could perhaps be broken down into four distinct but overlapping sub-processes.

(1) Production of a filament by apical growth

(2) Transverse division of the segments after a resting period

(3) Division of the daughter-cells longitudinally

(4) Development of the daughter and granddaughter cells as branches following the same pattern or development as the main axis

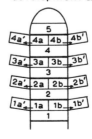

Fig. 5.5

101

In stage 4, '1a'' would be likely to appear before '1b'' and in Fritsch's illustration (F.264) which is less diagrammatic there are distinct signs of one side-branch lagging behind its opposite number as the process continues.

This development, which may or may not be a true picture of events, is characterised by the fact that each initial cell undergoes two important divisions which help to determine the general structure of the plant. This brings me to the second part of my hypothesis.

Examples of Fibonacci numbers often seem to be derived from some modified form of binary division. Modifications mentioned so far are:

rabbits waiting for a double period before the first delivery; drones having only one parent instead of two; committee members having their right to nominate new members restricted to exactly two nominations.

I suggest that the Fibonacci numbers often observed in the spiral development of plants may come from a modified form of cell division. One possible modification in the development of a ring of initials from one single initial is that there is a time-lag before a newly created cell joins in the process, as in the rabbit example. A more likely arrangement is that cells drop out of the process after two generations, following the same rule as the committee members. This would not preclude them from further divisions later on when the time comes to initiate a series of primordia. In either case the resulting arrangement of initials would be as depicted in one of the committee diagrams of fig. 5.2 a, or one with more individual cells as appropriate.

The diagram opposite is taken from F.J. Richards, 'Phyllotaxis; its quantitative expression and relation to growth in the apex', in *Philosophical Transactions of the Royal Society of London, series B, Biological Sciences*, ———, 629 (1951) 512.

The diagram is stated to be an ideal arrangement consisting of a family of 8 equiangular spirals crossing another family of 5 and forming curved quadrilinear regions which represent primordia. These primordia have been numbered according to their distance from the central apex and so also according to age and size, number 1 being the oldest. The angular spacing of the first 21 regions is exactly the same as that of the diagram for the committee with 21 members. Four extra regions 22 to 25 also appear in the diagram. The bare apex in the centre suggests a star polygon of five points: any 8 consecutive regions also form an approximate star polygon, and so also for 13 and 21. The angle of divergence between 1 and 2 is approximately the Fibonacci angle $360° \times (1-f) = 137 \cdot 5°$, or else $360° \times f = 222 \cdot 5°$, according as you go the shorter or longer way round ($f = 0 \cdot 6180$).

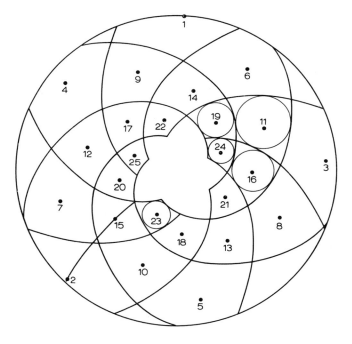

Fig. 5.6

If by any chance the hypothesis of an age-label combined with two special divisions is correct the sequence of cell division would be:

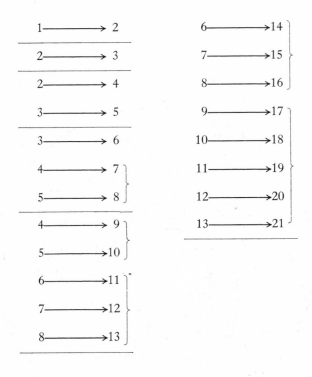

5.6   The origin of the spiral pattern

The circles in fig. 5.6 drawn by Richards provide a probable explanation of the spiral pattern. If the subsequent history of each initial in the ring is that it proceeds to form a circular disc by further division of itself and its new derivatives, and if expansion is at a uniform rate, the size of the various discs would vary with the age of the initial and its final shape would be conditioned by pressure from neighbouring discs. The dot in each region may be regarded as the growth centre of the disc and it is evident that the curve through the growth centres 6, 11, 16, 21, say, follows the same spiral path as the family of 8 spirals in the diagram. However Richards pointed out that other ideal curves, for example the ones below, can be drawn based on exactly the same points.

104

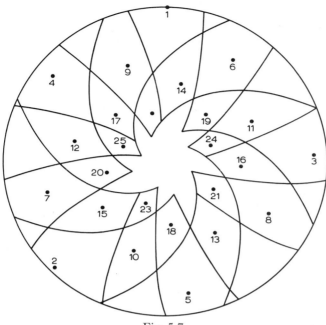

Fig. 5.7

The shape of the primordia in the new arrangement suggests that radial growth has outstripped tangential growth so that although the pattern as a whole is expanding at exactly the same rate the territory associated with each growth centre is different in shape while being of the same area. If tangential growth took precedence the shape of a primordium would be more like

Fig. 5.8

leading to a tightly folded system such as that of a rosebud. A typical spiral of this sort would be through the growth centres 1, 3, 5, 7, 9, . . .

It was said in section 4.16 that rotation of a spiral produced much the same effect as uniform expansion from the pole. Fig. 5.6 can be used to verify this. If the reader will make separate tracings of the two families of spirals and superpose them to form the diagram of Fig. 5.6 then rotation of the upper paper about the centre in one direction will produce the effect of uniform expansion, and in the other will simulate contraction.

105

5.7 Other Fibonacci systems

In an appendix to his paper Richards gave the observed facts for plants which follow other Fibonacci sequences. In such cases the divergence angle between successive primordia varies according to the sequence followed:

System	Divergence angle
1, 2, 3, 5, 8 . . .	137·5°
3, 4, 7, 11, 18 . . .	99·5°
4, 5, 9, 14, 23 . . .	78·0°
5, 6, 11, 17, 28 . . .	64·1°
5, 7, 12, 19, 31 . . .	151·1°

The hypothesis of section 5.5 provides a plausible explanation of these angles. As an example take the sequence 3, 4, 7, 11, 18, . . . The order of cell division up to 11 can be guessed by working backwards using the same pattern as before. The last stage is likely to be

The diagrams below give all the possible star polygons having 11 points.

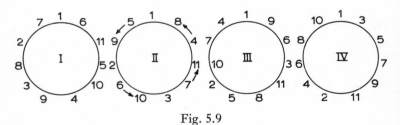

Fig. 5.9

Only the arrangement in II meets the requirement so we choose this. The stage before can be obtained by eliminating 8, 9, 10, 11. Then 5, 6, 7 must be deleted.

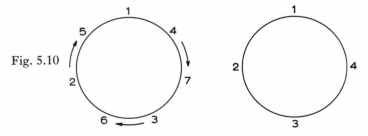

Fig. 5.10

106

The sequence of cell division seems to be:

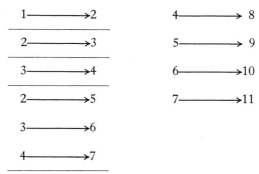

The angle of divergence is now derived from observing the position of 2 at successive stages. Its position is determined by the sequence 1/4, 2/7, 3/11, 5/18, the numerators being in the sequence 0, 1, 1, 2, 3, 5, 8, . . . and the denominators in the sequence 1, 3, 4, 7, 11, . . . Successive values for these ratios are: 0·250, 0·286, 0·273, 0·278 which suggest that they converge to a definite value. To find this value compare corresponding terms of the two derived geometric progressions each with common ratio $(f+1)$:

$1, f+1, f+2, 2f+3, 3f+5, \ldots$
$f+3, 3f+4, 4f+7, 7f+11, 11f+18, \ldots$ (see Section 4.5).

Corresponding terms of these series are in the ratio $1/(f+3)$ and it may be presumed that $3/11, 5/18 \ldots$ tend to this limit.

Now, $1/(f+3)=0·2765$,

hence the ideal divergence angle for this series is

$$1/(f+3) \times 360° = 99·50°.$$

which agrees with Richards' observed angle given in section 5.7.

In the same way the arrangement up to 12 in the sequence 5, 7, 12 may be deduced:

Fig. 5.11

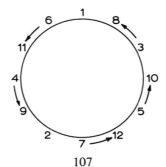

The angle of divergence turns out to be $1/(3-f) \times 360° = 151 \cdot 1°$. The other sequences also lead to the divergence angles observed by Richards which is a further indication that the hypothesis may be valid.

5.8 Variation within a sequence

Why do some plants divide 5 to 8 and others 21 to 34? One obvious factor would be the number of generations of initials. Another factor might be the rate of expansion. In pine-cones, for example, the fatter cones have 13–8 or 8–5 spirals while the long thin cones have 5–3 spirals, possibly because of a higher rate of longitudinal growth.

The measure of expansion usually taken is the 'plastochrone ratio', i.e. the ratio of the radial distances of two successive primordia from the centre of the apex. Richards gives the following table for plants in the 1, 2, 3, 5 . . . sequence:

?	Plastochrone ratio
3– 5	1·204
5– 8	1·073
8–13	1·027
13–21	1·010
21–34	1·004
34–55	1·0015

This table, showing that slower expansion goes with a higher number of spirals, indicates that rate of expansion is an important factor in determining the number to be seen.

5.9 Bijugate arrangements in Fibonacci spirals

Richards' paper also mentioned cases of spiral phyllotaxis with Fibonacci angles in a bijugate system, i.e. when leaves occur in pairs opposite each other. Such an arrangement might result from a slight modification of the development already discussed. If two initials 1 and 1' began simultaneously to develop further initials in a ring round the apex this would be the result.

Two interlacing spirals would be seen, 1, 2, 3′, 4′, 5 and 1′, 2′, 3, 4, 5′, with pairs of leaves (1, 1′), (2, 2′), . . . from the bottom of the stem towards the top.

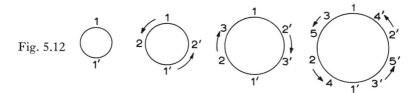

Fig. 5.12

5.10 Stages of development

To return to the main hypothesis: it may be conjectured that for the plants which display Fibonacci patterns the whole process takes place in three stages which might be distinct or overlapping.

(i) Development of the distal zone by repeated mitoses until there is a sufficiently large apex to support the next stage.

(ii) Transition of the apex to a relatively quiescent state and the development of the initiating ring in the sub-distal zone from a single initial.

(iii) After completion of the ring, or earlier, development of the initials in it, one by one, to form primordia, the process becoming evident in the organogenic zone. One more quotation, this time from C. Leopold, *Plant Growth and Development*, p.8, is appropriate here. In describing a recent experiment the author says:

'A striking illustration of the mutual effects of cells on differentiation can be seen in the development of a simple meristem. In a tissue culture, after a colony of cells has reached a certain size there appear localised organised meristems. . . . The first step in the emergence of the meristem seems to be the stimulation of one cell in the colony to commence divisions, and this apparently occurs immediately adjacent to a lignified cell – a simple vascular element – suggesting that the latter is providing some stimulus to cell division. More lignified cells are then laid down near the dividing cell, suggesting that the dividing cells somehow encourage the formation of the lignified or vascular elements. This process is multiplied until there emerges finally a dome of meristem. . . . In this tissue-culture situation, we seem to be observing a recapitulation of the basic steps in the emergence of a tissue, with conspicuous implications of chemical messengers passing between the cells and regulating the sequence.'

5.11 Summary of suggestions

To conclude the chapter a brief summary of my conjectures may be useful: perhaps some if not all of them may be true or may make some small contribution to the understanding of underlying machinery of Fibonacci patterns in nature,

(a) That Fibonacci numbers, because of their widespread occurrence and their known connections with modified forms of binary fission may possibly originate with cell-division.

(b) That a cell once formed may retain its individuality permanently, carrying within it an age-label which may affect its response to various stimuli.

(c) That one possible modification is the restriction of a particular phase of cell-division to two generations per cell.

Bibliography for Chapter 5.

N.J. Berrill, *Growth, Development and Pattern* (1961)

F.A.L. Clowes, *Apical Meristems* (1961)

E.J.H. Corner, *The Life of Plants* (1964)

K. Esau, *Plant Anatomy* 2nd ed. (1965)

F.E. Fritsch, *The Structure and Reproduction of the Algae II* (1952)

A.C. Leopold, *Plant Growth and Development* (1964)

C. Needham, *The Growth Process in Animals* (1964)

F.J. Richards, 'Phyllotaxis, its quantitative expression and relation to growth in the apex' in *Royal Society of London, Series B, Biological Sciences*, Vol. 235, No. 629 (1951)

Strasburger's Textbook of Botany New English ed., rewritten by R. Harder, W. Schumacher, F. Firbas, D. von Denffer, translated by P. Bell, D. Coombe (1965)

C.W. Wardlaw, *Organisation and Evolution of Plants* (1965)

P.B. Weisz and M.S. Fuller, *The Science of Botany* (1962)

6

NEW GROUPS FROM OLD AND GROUPS WITHIN GROUPS

This chapter has two main themes and one subsidiary, all inter-related. Section 1–7 deal with direct product groups, sections 8–12 with quotient groups, and section 13 with equivalence classes. Quotient groups were first mentioned on an experimental basis in section 3.4. Those introductory ideas are again used in the first portion of this chapter. A more detailed and rigorous account of quotient groups follows in the second portion. The reader who prefers not to postpone the rigorous approach could read sections 8–12 before sections 1–7. The final section of the chapter is an elementary treatment of equivalence, for the benefit of those readers who may not have previously met this important concept in its general algebraic sense; it underlies the discussion of modular arithmetic in section 2.1 and is essential to the understanding of the rôle of cosets in quotient groups which is developed in the next chapter.

6.1 The direct-product of two groups

It is always possible to derive from any two groups J and K a new and larger group, by combining them in a particular way. The new group is called the *direct product* of J and K, is written $J \times K$ and is obtained as follows:

(a) Its elements are the complete set of pairs (j, k) where j is an element of J and k is an element of K. If J is of order m and K of order n then $J \times K$ is of order mn. The pairs are not ordered; (k, j) is merely another way of writing (j, k). However it is a convenient notation consistently to put the j-element first and the k-element second, in making any tabulation.

(b) In the new group the product of any two pairs (j_1, k_1) and (j_2, k_2) is the pair $(j_1 j_2, k_1 k_2)$ where $j_1 j_2$ is the product according to the rule of combination for J and $k_1 k_2$ is the product according to the rule of combination for K. We shall take $\boxed{D}$ as the symbol for the direct product rule of combination.

111

6.2 An example of a direct product group

As a first example let us find the direct product $J \times K$ when J is of type C_3 and K of type C_2. J and K may be specified by their tables, using symbols e and f for their respective identity elements.

	J		
	e	a	b
e	e	a	b
a	a	b	e
b	b	e	a

	K	
	f	g
f	f	g
g	g	f

The elements of $J \times K$ are then (e,f), (a,f), (b,f), (e,g), (a,g), (b,g). It is convenient to follow this order when setting out the multiplication table for $J \times K$.

Table 6.1

	(e,f)	(a,f)	(b,f)	(e,g)	(a,g)	(b,g)
(e,f)	(e,f)	(a,f)	(b,f)	(e,g)	(a,g)	(b,g)
(a,f)	(a,f)	(b,f)	(e,f)	(a,g)	(b,g)	(e,g)
(b,f)	(b,f)	(e,f)	(a,f)	(b,g)	(e,g)	(a,g)
(e,g)	(e,g)	(a,g)	(b,g)	(e,f)	(a,f)	(b,f)
(a,g)	(a,g)	(b,g)	(e,g)	(a,f)	(b,f)	(e,f)
(b,g)	(b,g)	(e,g)	(a,g)	(b,f)	(e,f)	(a,f)

The very regular appearance of the table is a characteristic of all direct product groups when arranged in an appropriate order and the following significant features may be observed.

(i) The top left-hand compartment displays the elements $(e,f),(a,f)$, (b,f) as a normal subgroup isomorphic with J.

(ii) The quotient group is isomorphic with K.

(iii) Not only in the top left-hand compartment but also in all the others the pairs exhibit the elements of J according to the pattern of J while an element of K is constant throughout.

112

(iv) The pattern of compartments is:

F	G
G	F

and in group $J \times K$ both compartments marked F are identical and both marked G are identical, whereas in many groups arranged on a normal subgroup two compartments containing the same set of elements may show considerable divergence in detail (see table 3.1 for example).

The new group we have constructed might be called $C_3 \times C_2$ or $C_2 \times C_3$: the order does not matter because each of the constituent groups has the same rôle in the direct product group. However $C_3 \times C_2$ is isomorphic with the group already known to us as C_6, a statement which the reader can easily verify by turning back to table 2.4 and comparing it with table 6.1 above. Alternatively it may be seen by taking the pair (a,g) as a generating element: starting from the identity element (e,f), (a,g) generates the sequence

$$(e,f) \; (a,g) \; (b,f) \; (e,g) \; (a,f) \; (b,g) \; (e,f). \; \ldots$$

6.3 Satisfaction of the group axioms

The group axioms should be verified for $J \times K$.

(a) Closure: if (j_1,k_1) and (j_2,k_2) are any two elements of $J \times K$ their product by definition is $(j_1 j_2, \; k_1 k_2)$. As $j_1,j_2 =$ some element of J and $k_1,k_2 =$ some element of K the product must be an element of $J \times K$.

(b) Identity element: the identity element of $J \times K$ is (e,f) where e is the identity element of J and f is the identity element of K.

(c) Inverses: the inverse of (j,k) is (j^{-1}, k^{-1}) each of the pair being the inverse element in the appropriate group.

(d) Associativity: from the rule of combination for $J \times K$ the associativity of its elements is assured because of the associativity necessarily possessed by elements of J and K.

6.4 The direct product $C_4 \times C_2$

Our second example is $C_4 \times C_2$, very similar to the previous one, but this time we shall use the notation of multiplication groups for both J and K as suggested in section 3.1.

$$J = \{1, a, a^2, a^3\} \qquad K = \{1, b\}$$

113

With this notation both identity elements have the same symbol, but we can distinguish between them in the direct product by regarding 1 in the first position of a pair as the identity element for J and in the second position as the identity element for K. The table for $J \times K$ is indicated below without every detail being filled in.

Table 6.2

	$(1,1)$	$(a,1)$	$(a^2,1)$	$(a^3,1)$	$(1,b)$	(a,b)	(a^2,b)	(a^3,b)
$(1,1)$	$(1,1)$	$(a,1)$	$(a^2,1)$	$(a^3,1)$	$(1,b)$	(a,b)	(a^2,b)	(a^3,b)
$(a,1)$	$\ldots$	$\ldots$	$\ldots$	$\ldots$				
$(a^2,1)$	$\ldots$	$\ldots$	$\ldots$	$\ldots$				
$(a^3,1)$	$\ldots$	$\ldots$	$\ldots$	$\ldots$				
$1,b$	$(1,b)$	(a,b)	(a^2,b)	(a^3,b)	$(1,1)$	$(a,1)$	$(a^2,1)$	$(a^3,1)$
a,b								
a^2,b								
a^3,b								

(The diagonals indicate the usual cyclic pattern of the items in the table not displayed in full.)

It will help us to identify this group among those already encountered in Chapter 3 if we examine the order of each of the elements. Four of them $(a,1)$ $(a^3,1)$ (a,b) and (a^3,b) are of order 4 while $(a^2,1)$ $(1,b)$ and (a^2,b) are of order 2. These considerations rule out the possibility of the group being C_8 (which has one element of order 2), or D_4 (which has five elements of order 2). The group whose table is given in section 3.12 fills the bill as regards order of elements, and as arranged in table 3.9 it is seen to have the same structure as table 6.2. Further if we consider the geometrical illustration which follows table 3.9 we see that the eight transformations of that group can be put into the form of pairs from two cyclic groups. If r stands for a rotation of 90° and m for reflection

114

in a horizontal mirror, then the transformations can be concisely expressed as follows:

$$f=(1,1) \qquad h=(1,m)$$
$$a=(r,1) \qquad g=(r,m)$$
$$d=(r^2,1) \qquad c=(r^2,m)$$
$$e=(r^3,1) \qquad b=(r^3,m).$$

Exercises

6.1   Are any or all of the following statements true?

(a) $C_5 \times C_2$ is isomorphic with C_{10},

(b) $C_6 \times C_2$ is isomorphic with C_{12},

(c) $C_4 \times C_3$ is isomorphic with C_{12}.

6.2   There is no difficulty about extending the idea of direct products to three constituent groups A,B,C. The elements of the direct product are all possible triples from the groups e.g. (a_1,b_1,c_1), (a_2,b_2,c_1). The product of the pair would be the triple $(a_1a_2,b_1b_2,c_1{}^2)$.

(a) Develop the 'multiplication' table for the direct product of groups A,B,C below and compare your result with the examples given in section 3.11.

A			B			C		
	1	a		1	b		1	c
1	a	1	1	1	b	1	1	c
a	1	a	b	b	1	c	c	1

(b) The group developed in (a) is of type $C_2 \times C_2 \times C_2$. Investigate the group $D_2 \times C_2$ (D_2 is Klein's group) and decide whether it is isomorphic with $C_2 \times C_2 \times C_2$ or $C_4 \times C_2$.

6.5 Commutativity in direct product groups

The rule of combination for direct product groups ensures that $J \times K$ is commutative if both J and K are commutative. It may happen, however, that one or both are non-commutative and as our last example we shall examine $D_3 \times C_3$. First we set out their tables:

J

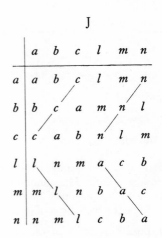

K

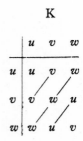

Next we set out a skeleton table for their direct product.

Table 6.3

	au	bu	cu	lu	mu	nu	av	bv	cv	lv	mv	nv	aw	bw	cw	lw	mw	nw
au	au	bu	cu	lu	mu	nu	av	bv	cv	lv	mv	nv	aw	bw	cw	lw	mw	nw
bu	bu	cu	au	mu	nu	lu	bv						bw					
cu	cu	au	bu	nu	lu	mu	cv						cw					
lu	lu	nu	mu	au	cu	bu	lv			V			lw			W		
mu	mu	lu	nu	bu	au	cu	mv						mw					
nu	nu	mu	lu	cu	bu	au	nv						nw					
av	av	bv	cv	lv	mv	nv	aw						au					
bv	bv																	
cv	cv																	
lv	lv			V						W						U		
mv	mv																	
nv	nv																	
aw	aw	bw	cw	lw	mw	nw	au						av					
bw	bw																	
cw	cw																	
lw	lw			W						U						V		
mw	mw																	
nw	nw																	

(To save space (a,u) is printed as au)

117

The table is intended to show the important features of the group:

(i) In each of the compartments marked U,V or W, the J-members of the pairs follow the pattern of the J group, while the K-members have a constant value u, v or w.

(ii) Compartments U, V, W are named to correspond with u, v, w. Each contains only six distinct elements. Those in U form a normal subgroup J⁕ isomorphic with J. Its quotient group, shown by the pattern of compartments, is isomorphic with K.

(iii) The three compartments marked U are identical and so are those marked V and those marked W.

(iv) The elements (a,u), (a,v), and (a,w) form a subgroup K⁕ of type C_3 also isomorphic with K.

(v) The group as a whole is non-commutative because J⁕ is non-commutative, but the elements of J⁕ and K⁕ commute with each other as can be seen from the rows and columns printed in bold. As this is the key fact about direct product groups an enquiry into the reason for it is relevant at this point. To take a typical example (b,u) is seen to commute with (a,v):

$$(b,u) \; \boxed{\text{D}} \; (a,v) = (ba, uv) = (b,v)$$
$$(a,v) \; \boxed{\text{D}} \; (b,u) = (ab, vu) = (b,v),$$

since a and u are identity elements in J and K.

(vi) The elements of J⁕ multiplied in turn by the elements of K⁕ produce all the elements of the group.

6.6 Conditions for any group to be isomorphic with a direct product group

There are five groups of order 8. Of these four contain a normal subgroup of type C_4 with a quotient group of type C_2; but only one of these is isomorphic with the direct product group $C_4 \times C_2$. How can we distinguish between them? If a group can be analysed into a normal subgroup A with quotient group B it is desirable to know what additional properties it must possess for us to be able to classify it as $A \times B$. Let us look again at table 6.3 but this time let us disregard its origin as a direct product group and consider it simply as an abstract group G. To assist us in this exercise let us re-name its elements with single letters $a,b,c, \ldots$ as indicated in the table below.

118

	a	b	c	d	e	f	h	j	k	l	m	n	p	q	r	s	t	u
a	**a**	**b**	**c**	**d**	**e**	**f**	**h**						**p**					
b	**b**	**c**	**a**	**e**	**f**	**d**	**j**						**q**					
c	**c**	**a**	**b**	**f**	**d**	**e**	**k**			H			**r**			P		
d	**d**	**f**	**e**	**a**	**c**	**b**	**l**						**s**					
e	**e**	**d**	**f**	**b**	**a**	**c**	**m**						**t**					
f	**f**	**e**	**d**	**c**	**b**	**a**	**n**						**u**					
h	**h**	**j**	**k**	**l**	**m**	**n**	**p**						**a**					
j																		
k			H							P						A		
l																		
m																		
n																		
p	**p**	**q**	**r**	**s**	**t**	**u**	a						**h**					
q			P							A						H		
r																		
s																		
t																		
u																		

(The large compartment labels span the nine 6×6 blocks in the pattern: top band **A H P**, middle band **H P A**, bottom band **P A H**. The large **A** is printed across the centre of the top-left block.)

In what follows the symbols A, H, P are taken, somewhat loosely, to represent sometimes whole compartments in the table and sometimes the cosets whose elements are in the compartments. The context will enable the reader to distinguish between the two uses.

If we were to show this table, with all its details filled in to someone who knew nothing about its previous history as a direct product group, he would be able to make the following statements about the structure of the group:

(i) It has a normal subgroup A with elements a,b,c,d,e,f of type D_3.

(ii) From the pattern of the compartments A,H,P the quotient group is of type C_3.

(iii) Compartments marked A are identical and so are those marked H and those marked P.

(iv) The elements a,h,p form a subgroup B also of type C_3 and isomorphic with the quotient group.

(v) Every element of A commutes with every element of B as indicated by the rows and columns printed in bold.

(vi) The elements of A multiplied in turn by the elements of B produce all the elements of the group.

All these observations of the internal structure of the group might lead our imaginary friend to suspect that he was looking at a group isomorphic with the direct product group $D_3 \times C_3$.

We are now in a position to state and prove the general conditions for a group G with subgroups A and B to be isomorphic with the direct product group $A \times B$. There are two conditions:

1. That every element of A commutes with every element of B.

2. That the elements of A multiplied in turn by the elements of B produce all the elements of G, each element occurring just once.

It is these two conditions which lead to the sets of identical compartments which are such a striking characteristic of all direct product groups. The formal proof follows:

Let $\maltese$ stand for the rule of combination in G and hence also for the rule in A and B, while $\boxed{D}$ stands for the direct product rule.

Condition (2) ensures that every element of G can be expressed in one way only as the product $a \maltese b$ of an element a from A and an element b from B. This means that every element of G can be matched in a one-to-one correspondence with a pair (a,b).

$$(a \maltese b) \longleftrightarrow (a,b) \tag{6.1}$$

For the isomorphism between G and $A \times B$ to exist it is necessary to show that for all $a_1 a_2 b_1 b_2$

$$(a_1 \maltese b_1) \maltese (a_2 \maltese b_2) \longleftrightarrow (a_1,b_1) \boxed{D} (a_2,b_2)$$
i.e. $(a_1 \maltese b_1) \maltese (a_2 \maltese b_2) \longleftrightarrow (a_1 \maltese a_2, b_1 \maltese b_2)$

Now $(a_1 \divideontimes b_1) \divideontimes (a_2 \divideontimes b_2)$ $=$ $a_1 \divideontimes (b_1 \divideontimes a_2) \divideontimes b_2$, associative property,

$=$ $a_1 \divideontimes (a_2 \divideontimes b_1) \divideontimes b_2$, condition (1),

$=$ $(a_1 \divideontimes a_2) \divideontimes (b_1 \divideontimes b_2)$, associative property

and from (6.1) above the last product is known to correspond to the pair $(a_1 \divideontimes a_2, b_1 \divideontimes b_2)$, so the isomorphism of G with $A \times B$ exists.

6.7 An application of the isomorphism

Conditions (1) and (2) of the last section as criteria for the isomorphism between G and $A \times B$ can be used with advantage in the following example.

Consider the multiplication group formed by eight of the residue classes modulo 20:

$$1, 3, 7, 9, 11, 13, 17, 19.$$

There is a cyclic subgroup A of type C_4 consisting of 1, 3, 9, 7 and another one B of type C_2 consisting of 1, 11. Each residue can be expressed as the product of two elements one from each subgroup:

$$1 = 1 \times 1 \qquad\qquad 11 = 1 \times 11$$
$$3 = 3 \times 1 \qquad\qquad 13 = 3 \times 11$$
$$9 = 9 \times 1 \qquad\qquad 19 = 9 \times 11$$
$$7 = 7 \times 1 \qquad\qquad 17 = 7 \times 11$$

The group is commutative and can therefore be identified as isomorphic with $C_4 \times C_2$. Its whole table can be written down at once by rote.

$\times$	1	3	9	7	11	13	19	17
1	1	3	9	7	11	13	19	17
$\cdot$ 3	3	9	7	1	13	19	17	11
9	9	7	1	3	19	17	11	13
7	7	1	3	9	17	11	13	19
11	11	13	19	17	1	3	9	7
13	13	19	17	11	3	9	7	1
19	19	17	11	13	9	7	1	3
17	17	11	13	19	7	1	3	9

6.8 Groups within groups

The idea of a subgroup is already familiar and in section 3.4 we observed that certain subgroups were particularly effective in organising the main group into cosets which were themselves the elements of a group. A subgroup which possesses this property is called a normal subgroup and the group of cosets is said to be its quotient group. We used these ideas freely while analysing direct product groups but it is now time to examine them more thoroughly. We need a precise definition of a normal subgroup from which we can establish the fact that a corresponding quotient group always exists.

6.9 The organising effect of a normal subgroup

Let us again consider a particular case of a subgroup which possesses the special property mentioned in the previous section. The multiplication table of a group of order 12 is fully set out in exercise 3.5. It can be organised by means of a subgroup $N=\{l,b,g,h\}$ the other two cosets being $P=\{a,e,m,f\}$ and $Q=\{c,d,k,j\}$.
This is the revised table:

Table 6.4

	✳	l	b	g	h	a	e	m	f	c	d	k	j
		N				**P**				**Q**			
N	l	l	b	g	h	a	e	m	f	c	d	k	j
	b	b	l	h	g	e	a	f	m	d	c	j	k
	g	g	h	l	b	m	f	a	e	k	j	c	d
	h	h	g	b	l	f	m	e	a	j	k	d	c
			N²				NP				NQ		
P	a	a	m	f	l	c	k	j	d	l	g	b	h
	l	l	f	m	a	d	j	k	c	b	h	g	l
	m	m	a	l	f	k	c	d	j	g	l	b	h
	f	f	l	a	m	j	d	c	k	h	b	l	g
			PN				P²				PQ		
Q	c	c	j	d	k	l	h	b	g	a	f	e	m
	d	d	k	c	j	b	g	l	h	e	m	a	f
	k	k	d	j	c	g	b	h	l	m	e	f	a
	j	j	c	k	d	h	l	g	b	f	a	m	e
			QN				QP				Q²		

There are two significant features to be noticed. First, the compartment marked PQ consists of all such products as $p \mathbin{*} q$ where p is an element of P and q is an element of Q; and QP consists of all such products as $q \mathbin{*} p$. Second, compartments PQ, QP, and N^2 resemble each other in containing only the four elements of the subgroup N but the compartments are not identical; the sixteen products are differently disposed in the three cases. In this respect N can be contrasted with a component subgroup of a direct product group which always produces sets of identical compartments.

6.10   The quotient group of N

Although we have not yet arrived logically at a precise definition of a normal subgroup we shall at this point decide that N is a normal subgroup on the grounds that it is a good organiser. Not every subgroup of the group possesses this organising property and the reader may like to investigate the subgroup $\{l,a,c\}$ from this angle.

To establish the existence of a quotient group corresponding with N several steps are necessary:

(a) We must define its elements.

(b) We must formulate a rule of combination which will enable us to specify the product of two elements.

(c) We must verify that the elements and the rule of combination together obey the group axioms.

Let us take these steps in order.

The elements of the new group F are the three cosets $N=\{l,b,g,h\}$, $P=\{a,e,m,f\}$, $Q=\{c,d,k,j\}$.

To specify the rule of combination requires some care. Following the procedure already used in Chapter 3 and in the first part of this chapter dealing with direct products we can write down the multiplication table for the three cosets:

Table 6.5

	N	P	Q
N	N	P	Q
P	P	Q	N
Q	Q	N	P

but this rather conflicts with our notation N^2, NP, NG, . . . used to label the compartments in Table 6.4. We need a rule of combination

123

© for multiplying cosets which will lead us to say that N^2 now interpreted as $N © N$ is N while $NP=N © P=P$ and $PQ=P © Q=N$. Let us abstract the portion of table 6.7 which leads us to convert NP into P.

		P			
		a	e	m	f
	l	a	e	m	f
	b	e	a	f	m
N	g	m	f	a	e
	h	f	m	e	a

The surprising fact which catches our attention is that all such products as $n \ast p$ (n from N, p from P), obtained by the rule of combination $\ast$ of the main group G, lead only to four distinct products a,e,m,f, which all belong to the same coset P. This suggests that our rule of combination © for any two cosets A,B of a group is to form all possible products $a \ast b$ and reduce the result to the set of *distinct* products

$$\{l, b, g, h\} © \{ a\ e\ m\ f\} = \{a, e, m, f\},$$

i.e. $\qquad$ N $\qquad$ © $\qquad$ P $\quad = \quad$ P .

As another example of the application of © to the cosets of another group it may be seen in tables 3.3 and 3.4 that

$$\{2, 4\} © \{5, 7\} = \{8, 6\},$$

while $\qquad \{2, 6\} © \{3, 7\} = \{4, 6, 8, 2\}.$

With this rule of combination it is readily seen that table 6.5 establishes the existence of a quotient group F of type C_3 corresponding to the normal subgroup N of G. The group F is of order 3 and it is evident from the way it is obtained that the order of F is equal to the order of G divided by the order of N.

6.11 Definition of a normal subgroup

So far we have been working on the assumption that the basic characteristic of a normal subgroup is that it is an exceptionally good organiser of a group. Any subgroup S of a group G has the effect of breaking up the group into cosets each containing s elements, and the group table can be arranged in compartments each containing s^2 elements. With a *normal* subgroup we have observed that the s^2 elements only contain s distinct elements. The cancellation property of G (see section 2.6) then ensures that the s elements appear once each in every line and column of any compartment.

We are now in a position to obtain another important property of normal subgroups and so to formulate a precise definition. A scrutiny of the way in which table 6.4 could be built up from the original table in exercise 3.5 provides the clue. We need not go very far in the process before coming to the necessary condition for N$=\{l,b,g,h\}$ to be a normal subgroup.

	*	l	b	g	h	a	e	m	f
			N				P		
N	l					a			
	b			N²		e		NP	
	g					m			
	h					f			
P	a	a							
	e	e		PN					
	m	m							
	f	f							

Step 1. Identify N$=\{l\ b\ g\ h\}$ as a subgroup with identity element l.

Step 2. Take any other element such as a and form the coset P$=\{a,e,m,f\}$ from the products $\{l,b,g,h\}*a$. This can be written P$=$N$*a$, the first column of compartment NP.

Step 3. Since l is the identity element the first column of compartment PN is also a,e,m,f.

Now if N is a normal subgroup (i.e. a good organiser) the first row of PN must also consist of the elements $\{a,e,m,f\}$ though not necessarily in that order. But the first row of PN is the set of products $a*\{l,b,g,h\}=a*$N. Hence a necessary condition for N to be a normal subgroup is that $a*$N$=$N$*a$, i.e. N must commute with a.

It follows that N must commute with every element in the group G. It certainly commutes with the elements $\{l,b,g,h\}$ because N is a subgroup, and as *a* was arbitrarily chosen in Step 2 it commutes with all the elements outside N.

Though ⓒ has just been defined as a special kind of multiplication between cosets we shall need the same symbol to cover multiplication between a coset and an element. Because of the cancellation property of any group, however, there is no difference between A ⓒ b and A✳b; and of course a ⓒ $b = a✳b$.

We have shown that the property of commuting with every element in the group is a necessary condition for a subgroup to earn the title of 'normal subgroup' and be an exceptionally good organiser. Another exactly equivalent way of stating the property is to say that the right cosets of a normal subgroup are always the same as the left cosets. For example we noticed in section 3.4 that the right cosets of the subgroup {1,5} were {2,6}, {3,7}, {4,8} while the left cosets were {2,8}, {3,7}, {4,6}. This showed that {1,5} was not a normal subgroup. From this point we shall consider the formal definition of a normal subgroup to be as follows:

N is a normal subgroup of a group G (✳) if

$$N✳g = g✳N,$$

for every element g of G.

Expressed even more symbolically this may be written shortly as

$$N \; \Delta G \text{ if } gN = Ng,$$

for all g $\in$ G.

6.12 Formal proof of the existence of the quotient group corresponding to any normal subgroup

Let N be a normal subgroup and A,B,C, . . . its cosets.

$$N = \{n_0, n_1, n_2, \ldots\}, n_0 \text{ being the identity element in G,}$$
$$A = Na \text{ (}a \text{ any element).}$$

Multiplication between cosets or between an element and a coset, e.g. N^2, NA, AB, aN, aB is to be understood as N ⓒ N, N ⓒ A, a ⓒ N, etc., the rule of combination ⓒ, being defined as in section 6.10.

To prove that N,A,B,C,...with rule of combination ⓒ, form a group F called the quotient group of N we verify that the system obeys a set of group axioms.

(a) *Associativity*

The rule of combination ⓒ being derived from the rule of combination of G is necessarily associative.

(b) *Closure*

For any X, Y in the set of cosets, N, A, B, . . .

$$XY = (Nx)(Ny), \quad x \text{ in } X, y \text{ in } Y,$$
$$= N(xN)y, \quad \text{associative property,}$$
$$= N(Nx)y, \quad \text{property of N,}$$
$$= N^2(xy), \quad \text{associative property,}$$
$$= N(xy), \quad \text{N a subgroup,}$$
$$= \text{coset of N, since } xy \text{ is an element of G.}$$

(c) *Identity element*

For any X in the set of cosets

$$NX = N(Nx) \qquad \text{and} \quad XN = (Nx)N$$
$$= N^2x, \text{ associative} \qquad = (xN)N, \text{ property of N,}$$
$$\qquad\qquad \text{property,} \qquad\qquad = xN^2$$
$$= Nx \qquad\qquad\qquad = xN$$
$$= X, \qquad\qquad\qquad = Nx, \text{ property of N,}$$
$$\qquad\qquad\qquad\qquad\qquad = X.$$

Hence N is the identity element in the set of cosets.

(d) *Inverse*

For any x in G

$$(Nx)(Nx^{-1}) = N(xN)x^{-1}$$
$$= N(Nx)x^{-1}$$
$$= N^2(xx^{-1})$$
$$= Nn_0, n_0 \text{ being the identity element in G,}$$
$$= N.$$

Hence every coset Nx has an inverse coset Nx^{-1} where $xx^{-1} = x^{-1}x = n_0$.

6.13 Equivalence classes

A housewife moves into a new house and begins to arrange her kitchen. She is likely to put all her teacloths in one drawer, because regardless of their shape, size, material or colour they all serve one purpose. Later on as space becomes a problem other things may find their way into this drawer and she may say to a friend who asks her where a tray cloth may be found 'it is in the same drawer as the teacloths.' In saying this the housewife is making use of an *equivalence relation* 'in the same drawer as' to define subsets of the set of objects which she likes to store in the kitchen drawers. This particular relation makes the following statements undeniably true:

1. The lemon-squeezer is in the same drawer as itself.

2. If the corkscrew is in the same drawer as the tin-opener then the tin-opener is in the same drawer as the corkscrew.

3. If the steel is in the same drawer as the carving-knife and the carving-knife in the same drawer as its fork then the steel is in the same drawer as the carving-fork.

These three properties which may or may not be possessed, severally or together, by *relations on sets* are called respectively the *reflexive*, *symmetric*, and *transitive* properties. They may be described symbolically as in the following definition of an equivalence relation.

If R is an equivalence relation on a set S and $a, b, c, \ldots \in S$ then:

     1. $a \, \mathrm{R} \, a$ (reflexive).

     2.   If $a \, \mathrm{R} \, b$ then $b \, \mathrm{R} \, a$ (symmetric).

     3.   If $a \, \mathrm{R} \, b$ and $b \, \mathrm{R} \, c$ then $a \, \mathrm{R} \, c$ (transitive).

Not all relations on sets are equivalence relations. Here are some which possess none or only some of the three properties:

Relation on set	Property
twin of (children in a family)	symmetric
mother of (people in a village)	none
shorter than (pencils in a box)	transitive
twice (rational numbers)	none
factor of (integers)	transitive
reciprocal of (rational numbers)	symmetric
inverse of (elements in a group)	symmetric
$\geqslant$ (real numbers)	reflexive and transitive

Relations are often expressed as equations: for example, the reciprocal relation between x and y where x and y are real numbers could be written $xy = 1$ and could also be pictured in the graph of a rectangular hyperbola if we chose to regard x and y as an ordered pair of Cartesian coordinates.

The most familiar example of an equivalence relation is that of equality. The old axiom 'things which are equal to the same thing are equal to one another' combines the symmetric and transitive properties.

Other examples of equivalence relations on sets are:

1. in the same class as (students in a school),
2. costs as much as (goods in a shop window),
3. as old as (children in a group),
4. congruent to (all triangles),
5. congruent to, modulo two (integers),
6. differs from by an even number (integers),
7. is parallel to.

In the last case it is possible to take the view that a line is not parallel to itself because there is no separation but if so the difficulty can be made to vanish by allowing the relation to read 'has the same direction as'. On the other hand everyone would agree that a line is not perpendicular to itself so that perpendicularity is not an equivalence relation; it has only the symmetric property. After considering this case it seems frivolous to reject the idea of a line being parallel to itself. Some delightful arguments can arise from such special examples of an equivalence relation. A seminar group discussing this topic was invited to consider whether 'brother of' is an equivalence relation on the set of boys in a school. The question drew two replies from one student, the first being, 'Nonsense, I can't be my own brother' and the second which followed immediately afterwards, 'It wasn't me who said that, it was my brother!' (See also Appendix IV).

The effect of an equivalence relation on a set is to partition it into disjoint subsets called equivalence classes, each containing at least one member. In the examples given above the equivalence classes were as follows:

1. school classes,
2. equally priced articles,
3. age groups,
4. triangles of the same shape and size,
5., 6. congruence classes – class 0 and class 1,
7. sets of lines having a common direction.

The idea of equivalence classes is very relevant to groups and has been introduced into this chapter because the effect of any subgroup is to partition the group into cosets which are equivalence classes. It does this in two ways, according to whether right or left multiplication is used to develop the cosets from the subgroup.

The equivalence relation on G might be described as 'in the same $\left\{\begin{matrix} \text{right} \\ \text{left} \end{matrix}\right\}$ coset as' but this is more descriptive of the result of the partitioning than the reason for it. A test frequently used to decide whether two elements a, b belong to the same right coset relative to a subgroup H is the condition

$$ab^{-1} = \text{some element of H,}$$
or more shortly $ab^{-1} \in$ H.

It can be proved that a, b are in the same right coset if and only if this condition is fulfilled. Assume first that $ab^{-1} = h$ where h is some element of H.

Then
$$\begin{aligned} (\text{H}a)b^{-1} &= \text{H}(ab^{-1}) \\ &= \text{H}h \\ &= \text{H.} \end{aligned}$$

Multiplication on the right by b then leads to
$$\text{H}a = \text{H}b,$$
i.e. a is in the same coset as b.

Conversely,

if $\qquad\qquad$ $\text{H}a = \text{H}b,$
Then $\qquad\quad$ $(\text{H}a)b^{-1} = (\text{H}b)b^{-1},$
i.e. $\qquad\qquad$ $\text{H}(ab^{-1}) = \text{H.}$

This means that
$\{e\, h_1\, h_2\, h_3 \ldots\}\,(ab^{-1}) = \text{H,}$
and hence that $e(ab^{-1}) = $ some element of H,
i.e. $\qquad\qquad$ $ab^{-1} = $ some element of H.

It can also be proved that $ab^{-1} = h$ is an equivalence relation on the set of elements in G. It is reflexive because $aa^{-1} = e = $ an element of H. It is symmetric because if ab^{-1} is in the subgroup H so is its inverse ba^{-1} (see section 2.8). And it is transitive because if ab^{-1} and bc^{-1} are in H so also is their product $(ab^{-1})(bc^{-1})$. Hence ac^{-1} is in H.

The corresponding equivalence relation which is a condition for a and b to belong to the same left coset is $a^{-1}b \in$ H and this of course implies that the inverse element $b^{-1}a$ is also in H. If the subgroup concerned is a normal subgroup N then all four of the conditions for a and b to be in the same coset may be assumed as fulfilled, i.e. ab^{-1}, ba^{-1}, $a^{-1}b$ and $b^{-1}a$ are all in N.

130

Exercises

6.3 Refer to the group whose table is given in exercise 3.5.

(a) Find the right and left cosets of the subgroup H={l,m,d}. Take any right coset R and find the set S of inverses of all the elements in R. Complete the tables RS and SR set out below

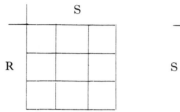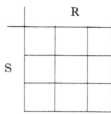

and verify that RS illustrates the fact that if a, b are in R then ab^{-1} is in H.

(b) Write out a suitable table which illustrates the property that if c, d are in a left coset L then $c^{-1}d$ is in H.

6.4 If a and b are two elements of a group G, a is defined as a conjugate of b if there is some element x in G such that $x^{-1}ax=b$.

(a) Show that the relation between a and b in this form is reflexive, symmetric, and transitive and so forms an equivalence relation.

(b) List the equivalence classes in the group of exercise 3.5 into which the relation of conjugacy partitions the group.

7

FIELDS, RINGS, AND HOMOMORPHISMS: ILLUSTRATIONS FROM THE FIBONACCI SEQUENCE

7.1 Fields

It becomes necessary to explain the technical term 'field' in order to be able to use it with precision. A field is a structure consisting of a set whose elements are related by two binary operations (i.e. rules of combination) as opposed to a group whose elements are related by one only. The elements are closed under both operations, usually addition and multiplication, and each operation has the commutative and associative properties. In symbols:

$$a+b=b+a \qquad ab=ba$$
$$(a+b)+c=a+(b+c) \qquad (ab)c=a(bc).$$

Also the two operations are connected by the distributive property for multiplication over addition:

$$a(b+c)=ab+ac.$$

There are two distinct identity elements, one for each operation, 0 for addition, 1 for multiplication:

$$a+0=0 \text{ and } a \times 1=1.$$

Each element has an inverse element a' (the negative of a) for addition, and each non-zero element has an inverse a^{-1} (the reciprocal of a) for multiplication:

$$a+a'=0 \text{ and } aa^{-1}=1.$$

The system has no zero divisors (see section 1.15).

If $$ab=0, \text{ then } a=0, \text{ or } b=0.$$

The last two properties mentioned (existence of a reciprocal and no zero divisors) are the distinguishing marks of a field when comparing it with other less well-behaved systems. The integers, for instance, do not possess reciprocals: the inverse of 3 according to the definition is $1/3$ but then $1/3$ is not an integer. The residues to a composite modulus do not all possess reciprocals and there are zero divisors among them.

The most familiar example of a field is the set of rational numbers p/q where p and q are integers and $q \neq 0$.

Other examples of fields are the set of real numbers, the set of numbers $a+b\sqrt{2}$ when a and b are rational numbers (a and b not both zero), and the set of residues to a prime modulus. Let us consider the residues modulo 5. All the usual rules for integers hold and in addition each non-zero element has a multiplicative inverse. The elements 2 and 3 are mutually inverse and the elements 1 and 4 are self-inverse. The usual symbol for this field is Z_5.

Another way of summing up the properties of a field is to say that the elements form a commutative group for addition and that the non-zero elements form another commutative group for multiplication, the two operations being linked together by the distributive property.

7.2 Rings

Now that we have defined a field it is profitable to contrast and compare it with a ring, which is another structure possessing two rules of combination. A ring resembles a field in that its elements form a commutative group for addition and it has the distributive property, but it may fail to qualify as a field for some or all of four reasons, all to do with multiplication:

1. Some of its elements may not possess a multiplicative inverse. This is the only reason for the integers not being classed as a field.

2. It may have zero divisors (see section 1.15). The set of residue classes to a composite modulus fails to qualify for this reason as well as for having elements with no multiplicative inverse.

3. The ring may not have a unity element. The even integers are an example.

4. It need not be commutative for multiplication. Matrices give us examples of non-commutative rings.

A field may be regarded as a particular example of a ring since it has all the properties required of a ring and some extra ones besides.

Let us now examine in some detail the ring Z_{10} of residue classes modulo 10. We already know that its elements form a commutative group for addition. Both $\{0, 5\}$ and $\{0, 2, 4, 6, 8\}$ are normal subgroups and it is instructive to set out both addition and multiplication tables of Z_{10} in an order based on $\{0, 5\}$. It will be seen that this subgroup and its cosets are also components of the multiplication table 7.2.

Table 7.1

+	0	5	1	6	2	7	3	8	4	9
0	0	5	1	6	2	7	3	8	4	9
5	5	0	6	1	7	2	8	3	9	4
1	1	6	2	7	3	8	4	9	5	0
6	6	1	7	2	8	3	9	4	0	5
2	2	7	3	8	4	9	5	0	6	1
7	7	2	8	3	9	4	0	5	1	6
3	3	8	4	9	5	0	6	1	7	2
8	8	3	9	4	0	5	1	6	2	7
4	4	9	5	0	6	1	7	2	8	3
9	9	4	0	5	1	6	2	7	3	8

Table 7.2

×	0	5	6	1	2	7	4	9	8	3
0	0	0	0	0	0	0	0	0	0	0
5	0	5	0	5	0	5	0	5	0	5
6	0	0	6	6	2	2	4	4	8	8
1	0	5	6	1	2	7	4	9	8	3
2	0	0	2	2	4	4	8	8	6	6
7	0	5	2	7	4	9	8	3	6	1
4	0	0	4	4	8	8	6	6	2	2
9	0	5	4	9	8	3	6	1	2	7
8	0	0	8	8	6	6	2	2	4	4
3	0	5	8	3	6	1	2	7	4	9

The following features of the two tables are of great interest: (i) The addition subgroup $\{0, 5\}$ is a good organiser for both tables, breaking them down into compartments containing the same set of pairs $\{0, 5\}$, $\{1, 6\}$, $\{2, 7\}$, $\{3, 8\}$ and $\{4, 9\}$. (ii) The pattern of the compartments is that of the field Z_5 whose tables are set out in section 1.3. The pattern inside the compartments is that of Z_2.

+	0	1
0	0	1
1	1	0

×	0	1
0	0	0
1	0	1

.

Exercise

7.1 Write out the addition and multiplication tables for Z_{10}, basing both on the subgroup $\{0, 2, 4, 6, 8\}$. What can be said about the structure of the compartments and the structure within the compartments?

As might be expected the tables for Z_8 reveal certain differences from those for Z_{10}.

Table 7.5

+	0	4	1	5	2	6	3	7
0	0	4	1	5	2	6	3	7
4	4	0	5	1	6	2	7	3
1	1	5	2	6	3	7	4	0
5	5	1	6	2	7	3	0	4
2	2	6	3	7	4	0	5	1
6	6	2	7	3	0	4	1	5
3	3	7	4	0	5	1	6	2
7	7	3	0	4	1	5	2	6

Table 7.6

×	0	4	1	5	2	6	3	7
0	0	0	0	0	0	0	0	0
4	0	0	4	4	0	0	4	4
1	0	4	1	5	2	6	3	7
5	0	4	5	1	2	6	7	3
2	0	0	2	2	4	4	6	6
6	0	0	6	6	4	4	2	2
3	0	4	3	7	6	2	1	5
7	0	4	7	3	6	2	5	1

The subgroup {0, 4} is again a good organiser since the same pairs {0, 4}, {1, 5}, {2, 6}, {3, 7} are apparent in both tables. However this time the pattern of compartments is not that of a field but of the ring Z_4 whose addition and multiplication tables are as follows:

+	0	1	2	3
0	0	1	2	3
1	1	2	3	0
2	2	3	0	1
3	3	0	1	2

×	0	1	2	3
0	0	0	0	0
1	0	1	2	3
2	0	2	0	2
3	0	3	2	1

There is no regular pattern to be seen within the compartments of the multiplication table of Z_8.

7.3 Ideals and quotient rings

From these examples it can be seen that a subgroup of the addition group in a ring plays a conspicuous part in organising the whole ring. Such a subgroup is called an ideal in the ring. One way of defining an ideal S in a ring R is by the following two conditions.
1.   S is a subgroup of R with respect to addition.
2.   Multiplication of any element in the ring, either on the left or on the right, by an element of the ideal, leads to an element in the ideal.

Since R is commutative for addition, by definition S is necessarily a normal subgroup. It can also be said that S partitions the ring into equivalence classes and that these classes have an addition and multiplication structure which is also a ring (and may in some cases be a field). By analogy with the quotient group this ring is called the quotient ring corresponding to S and may be symbolised by R/S. It is not proposed to furnish rigorous proof of these statements. Ideals and quotient rings have been introduced here by particular examples only for the reason that they have a bearing on the later subject matter of this chapter.

7.4 Homomorphisms

We have now collected a sufficient number of structures to be able to introduce the general idea which is called *homomorphism*. A homomorphism is defined to be *a mapping from one algebraic system to a like algebraic system which preserves structure*. In the second example of section 8.2 the ring Z_8 was mapped onto the ring Z_4 in a two-to-one correspondence.

$$Z_8 \qquad Z_4$$

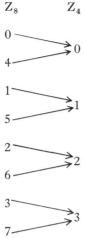

It is common practice to call the mapping φ and the image of any element a in Z_8 is then an element of Z_4 called $\varphi(a)$. For example $\varphi(0)=0$ and here the first zero belongs to Z_8 and the second to Z_4. Similarly $\varphi(2)=2$ and also $\varphi(6)=2$.

The important feature of the homomorphism is that the structure of the ring Z_8 is preserved by φ in respect to both addition and multiplication.

In symbols
$$\varphi(a)+\varphi(b)=\varphi(a+b),$$
$$\varphi(a)\times\varphi(b)=\varphi(a\times b).$$

This is not a new idea. We defined isomorphisms in a similar way in section 2.10. In fact an isomorphism may be regarded as a particular case of homomorphism with a one-to-one instead of a many-to-one correspondence between element and image.

Below is a particular example of preservation of the addition structure under the mapping set out above from the ring Z_8 to the ring Z_4.

$$\varphi$$

$$
\begin{aligned}
\varphi(5) &= 1, \\
\varphi(2) &= 2, \\
\varphi(5)+\varphi(2) &= 1+2, \\
&= 3, \\
\varphi(5+2) &= \varphi(7), \\
&= 3,
\end{aligned}
$$

$$
\begin{array}{cc}
Z_8 & Z_4 \\
0, 4 & \longrightarrow 0 \\
1, 5 & \longrightarrow 1 \\
2, 6 & \longrightarrow 2 \\
3, 7 & \longrightarrow 3
\end{array}
$$

hence
$$\varphi(5)+\varphi(2)=\varphi(5+2).$$

Here also is an example of preservation of the multiplicative structure:

$$\varphi(2)=2,$$
$$\varphi(6)=2,$$
$$\varphi(2)\times\varphi(6)=2\times2=4=0,\ \text{in}\ Z_4,$$
$$\varphi(2\times6)=\varphi(12)=\varphi(4\ \text{in}\ Z_8)=0\ \text{in}\ Z_4,$$

hence $\qquad \varphi(2)\times\varphi(6)=\varphi(2\times6).$

Tables 7.1 and 7.2 for Z_{10} also clearly display a two-to-one homomorphism from the ring Z_{10} onto the ring Z_5 (although Z_5 is a field it may also be described as a ring and we do so here because a homomorphism must be from an algebraic system to a like algebraic system). The mapping is $0, 5 \longrightarrow 0; 1, 6 \longrightarrow 1$ and so on.

A homomorphism may also be from a group onto a group as we have already seen in the additive structure of the ring. A most important case is one which we studied under another guise in chapter 6. If a group G has a normal subgroup N then there is a homomorphism from G to the quotient group G/N. The mapping is given by

$$a \longrightarrow Na,$$

where a is any element of G. We have already established this homomorphism under the heading 'closure' in section 6.12.

$$(Na)(Nb)=N(ab). \tag{7.1}$$

All we have to do is to recognise that in this case $\varphi(a)=Na$, so that equation (7.1) can be recast as

$$\varphi(a) \times \varphi(b)=\varphi(ab).$$

To take a particular case table 6.4 shows a homomorphism of this kind from the group A_4 of order 12 onto the group A_4/N of order 3, the detailed four-to-one mapping being

$$l, b, g, h \longrightarrow N$$
$$a, e, m, f \longrightarrow Na=P$$
$$c, d, k, j \longrightarrow Nc=R.$$

A homomophism might however be from a group G onto another group G which is not the quotient group of a normal subgroup of G; but in that case G is always isomorphic with the quotient group of some normal subgroup of G which is then called the *kernel* of the homomorphism. The diagram opposite (figure 7.1) may help to interpret these statements. It shows the relative disposition of the constituents of a homomorphism from a group G of order 6 onto a group $\overline{G}$ of order 2.

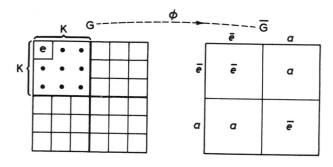

(Fig. 7.1)

The kernel K always contains e the identity element of G and consists of the elements forming a normal subgroup of G which are mapped onto e the identity element of $\overline{G}$. G and $\overline{G}$ may be quite independent. G could be the group of rotations of an equilateral triangle, $\overline{G}$ the residue classes mod 2 under addition.

If a homomorphism is from a ring R onto a ring $\overline{R}$ this also has a kernel which is an ideal of R. A formal definition of the kernel of this homomorphism is that it is the set of elements in R which is mapped by φ onto the zero element of $\overline{R}$. We have already seen in the mapping from Z_8 to Z_4 that $\{ 0, 4 \}$ is the set of elements in Z_8 which are mapped onto 0 in Z_4, so $\{ 0, 4 \}$ is the kernel of this particular homomorphism.

7.5   Groups, rings, and fields from the Fibonacci sequence

It may be recalled from the description in section 4.5 that this sequence, while obeying the Fibonacci rule that each term is the sum of the preceding pair, is also a sequence of powers of $f+1$. It can be made to yield a fine crop of commutative groups by using various prime moduli to reduce the terms. For example, making 1 the starting point with 3 as a modulus, the sequence recurs after 8 terms

$$1, f+1, f+2, 2f, 2, 2f+2, 2f+1, f, \ldots$$

With zero added there are nine elements which form a field. For addition they make a group $C_3 \times C_3$ constructed by adding in turn the elements of the cyclic sub-group $\{0, 1, 2\}$ to those of the cyclic sub-group $\{0, f, 2f\}$.

	0	1	2	f	$f+1$	$f+2$	$2f$	$2f+1$	$2f+2$
0	0	1	2	f	$f+1$	$f+2$	$2f$	$2f+1$	$2f+2$
1	1	2	0	$f+1$	$f+2$	f	$2f+1$	$2f+2$	$2f$
2	2	0	1	$f+2$	f	$f+1$	$2f+2$	$2f$	$2f+1$
f	f	$f+1$	$f+2$	$2f$	$2f+1$	$2f+2$	0	1	2
$f+1$	$f+1$	$f+2$	f	$2f+1$	$2f+2$	$2f$	1	2	0
$f+2$	$f+2$	f	$f+1$	$2f+2$	$2f$	$2f+1$	2	0	1
$2f$	$2f$	$2f+1$	$2f+2$	0	1	2	f	$f+1$	$f+2$
$2f+1$	$2f+1$	$2f+2$	$2f$	1	2	0	$f+1$	$f+2$	f
$2f+2$	$2f+2$	$2f$	$2f+1$	2	0	1	$f+2$	f	$f+1$

In the table, A=$\{0,1,2\}$ has been used as a normal subgroup and its quotient group is clearly C_3 since the compartments fall into the pattern seen below:

	A	B	C
A	A	B	C
B	B	C	A
C	C	A	B

The eight non-zero elements, being powers of the element $f+1$, must form the group C_8 for multiplication. We can also take it for granted that the distributive property $a(b+c)=ab+ac$ holds good for all choices of elements so that all the conditions for a field are fulfilled (see section 7.1).

The quickest way of obtaining the sequence of 8 terms is not to write out the algebraic sequence in full but to form the Fibonacci sequence of integers modulo 3 until recurrence shows that the cycle is complete:

$$0\ 1\ 1\ 2\ 0\ 2\ 2\ 1\ 0\ 1\ 1\ \ldots$$

Pairs of consecutive integers then give the sequence $1, f+1, f+2, \ldots$, the pair 1, 2 yielding $f+2$ and in general a pair a,b yielding $af+b$.

7.6 Modulus 7

Applying this technique with modulus 7 we get a sequence of 16:

$$0\ 1\ 1\ 2\ 3\ 5\ 1\ 6\ 0\ 6\ 6\ 5\ 4\ 2\ 6\ 1 \mid 0\ 1\ 1 \ldots$$

Two other sequences of sixteen elements may be obtained:

$$0\ 2\ 2\ 4\ 6\ 3\ 2\ 5\ 0\ 5\ 5\ 3\ 1\ 4\ 5\ 2 \mid 0,$$
$$0\ 3\ 3\ 6\ 2\ 1\ 3\ 4\ 0\ 4\ 4\ 1\ 5\ 6\ 4\ 3 \mid 0.$$

These sequences with zero added provide all the material for a field of 49 elements.

For addition the group is $(C_7 \times C_7)$ of the same type as $(C_3 \times C_3)$ shown above. For multiplication the first sequence provides a sub-group C_{16}.

$$A = 1, f+1, f+2, \ldots$$

Its two cosets are

$$B = 2, 2f+2, 2f+4, \ldots,$$
$$C = 3, 3f+3, 3f+6, \ldots$$

The quotient group is C_3. To verify this in any particular case the relationship $f^2 = 1 - f$ may be used (see section 4.6).

If $BC = A$ then the product of any pair of elements from B and C respectively should be an element of A,

e.g.
$$\begin{aligned}
(2f+2)(f+3) &= 2f^2 + 8f + 6 \\
&= 2f^2 + f + 6 \pmod{7} \\
&= 2(1-f) + f + 6 \\
&= 6f + 1, \text{ i.e. an element from A.}
\end{aligned}$$

Many other prime moduli among which may be included 23, 43, 97, fall into this pattern and provide straightforward fields based on cyclic subgroups. However, another set of moduli amongst which 11 is a typical example, yields a different structure which has some very interesting features.

7.7 Modulus 11

There are 11^2 elements which form the group $(C_{11} \times C_{11})$ for addition. For multiplication the non-zero elements break down into three distinct groups each with its own identity element. The basic Fibonacci

141

sequence runs out after 10 elements

$$0\ 1\ 1\ 2\ 3\ 5\ 8\ 2\ 10\ 1.$$

This leads to a subgroup of 10.

$$(1) \quad =\{1, f+1, f+2, \ldots\}.$$

Its nine cosets (2)—(10) are based on sequences as follows:

(2)	0	2	2	4	6	10	5	4	9	2,
(3)	0	3	3	6	9	4	2	6	8	3,
(4)	0	4	4	8	1	9	10	8	7	4,
(5)	0	5	5	10	4	3	7	10	6	5,
(6)	0	6	6	1	7	8	4	1	5	6,
(7)	0	7	7	3	10	2	1	3	4	7,
(8)	0	8	8	5	2	7	9	5	3	8,
(9)	0	9	9	7	5	1	6	7	2	9,
(10)	0	10	10	9	8	6	3	9	1	10.

Altogether these elements form a group G of order 100 for multiplication, the quotient group $G/(1)$ being C_{10}.

Exercise

7.2 Verify this statement and decide in which order the cosets should be placed so that the quotient group displays the characteristic diagonal pattern of a cyclic group.

There remain 20 elements arising from the three sequences

$$
\begin{aligned}
K &= 3,\ 2,\ 5,\ 7,\ 1,\ 8,\ 9,\ 6,\ 4, 10,\\
L &= 8, 10,\ 7,\ 6,\ 2,\\
M &= 1,\ 4,\ 5,\ 9,\ 3.
\end{aligned}
$$

K leads to one cyclic group for multiplication and L and M between them to another cyclic group. It is best to set them out fully:

Based on K	*Based on L, M*
$3f+2$ (identity element)	$8f+10$ (identity element)
$2f+5$	$5f+9$
$5f+7$	$10f+7$
$7f+1$	$9f+3$
$f+8$	$7f+6$
$8f+9$	$3f+1$
$9f+6$	$6f+2$
$6f+4$	$f+4$
$4f+10$	$2f+8$
$10f+3$	$4f+5$

142

The K group has been arranged in powers of $2f+5$ because with this element as generator the Fibonacci rule is seen to apply to successive elements; with the L, M group the best that can be done is to arrange it as powers of $5f+9$ in which case the Fibonacci rule applies to alternate elements.

For the operation of addition either set with zero forms a cyclic group and any non-zero element may be used as generator. Another way of saying this which can easily be checked is that the set consists of integral multiples 1–10 of any selected element. It follows that each of these small sets of 11 elements fulfils all the conditions for a field. This is not true of the set of 101 elements because it is not closed for addition: for example, f and 8 in G combine to make $f+8$ in K.

7.8 Multiplication between groups (modulus 11)

Appropriate names for the two minor groups are (0_1) and (0_2) because each possesses the following property akin to that of zero.

Multiplication of any element of (0_1) by an element of the major group G yields an element of (0_1). Both (0_1) and (0_2) play the rôle of a zero in relation to the quotient group of the cosets (1)–(10). Below is part of the multiplication table for these cosets arranged to illustrate the structure.

Table 7.3

	(0_1)	(1)	(2)	(4)	(8)	(5)	(10)	(9)	(7)	(3)	(6)
(0_1)	(0_1)	(0_1)	(0_1)	(0_1)	(0_1)	(0_1)	(0_1)	(0_1)	(0_1)	(0_1)	(0_1)
(1)	(0_1)	(1)	(2)	(4)	(8)	(5)	(10)	(9)	(7)	(3)	(6)
(2)	(0_1)	(2)	(4)	(8)	—	—	—	—	—	—	—
(3)	(0_1)	(4)	(8)	—	—	—	—	—	—	—	—

The set (0_1) in this table may be replaced by (0_2). There remains to be examined the product of (0_1) and (0_2). First consider the product of the identity elements

$$(3f+2)(8f+10)=2f^2+2f+9$$
$$=2(1-f)+2f+9$$
$$=0 \,(\text{mod } 11).$$

143

Any pair, one from each set, would give the same zero result since (0_1) consists of integral multiples of $(3f+2)$ and (0_2) of integral multiples of $(8f+10)$. Hence the table for $(0_1)(0_2)$ is

$\times$	(0_1)	(0_2)
(0_1)	(0_1)	0
(0_2)	0	(0_2)

and the system as a whole possesses zero divisors.

The two small sets (0_1) and (0_2) certainly possess some remarkable properties:

(i) Together with zero each forms a finite field.

(ii) The set (0_1) consists of a single Fibonacci sequence and (0_2) of two small sequences interwoven.

(iii) Any member of either set if multiplied by the integers 1–10 in turn produces the other members of the set.

(iv) The product of any pair of elements, one from each set, is zero.

(v) Each set is symmetrical in that each integer occurs twice and twice only.

(vi) The sum of the two identity elements for multiplication is unity.

On an empirical basis all these properties, appropriately modified where necessary as a result of changing the modulus, are found to hold for certain other prime moduli such as 29, 31, 41, 61.

7.9 Classification of the structure

How should the whole structure be classified? It is not a field since the non-zero elements as a whole do not form a multiplicative group. It does conform to the requirements of a ring and (0_1) and (0_2) with zero added both qualify as ideals in the ring since each is an additive group and any element of (0_1) multiplied by any element in the ring produces an element of (0_1), the same being true of (0_2).

What is particularly interesting about this ring is that the Fibonacci rule has organised it for multiplication in a completely different way from that described earlier for Z_{10} and Z_8. To bring out the difference let us look at Z_{15} and compare it with the Fibonacci ring. There is no difficulty in organising Z_{15} in the usual way on multiplies of 5 or 3. If we choose the latter then $\{0, 3, 6, 9, 12\}$ is an ideal and its additive cosets are $\{1, 4, 7, 10, 13\}$ and $\{2, 5, 8, 11, 14\}$. These cosets may be expected to have the structure of the field Z_3.

We shall now show that Z_{15} can be organised in a different way for multiplication to bring out its similarity with the Fibonacci ring.

(a) The fourteen non-zero elements may be broken down into three distinct groups each with a separate identity element:

$$0_1 = \{6, 3, 9, 12\}, \text{ identity element } 6,$$
$$0_2 = \{10, 5\}, \text{ identity element } 10,$$
$$G = \{1, 2, 4, 8, 7, 14, 13, 11\}, \text{ identity } 1.$$

(b) The sets (0_1) and (0_2) with zero added are subfields in the ring, because they form additive groups.

(c) The sets (0_1) and G combine to give a multiplication table having the same structure as the multiplication table of Z_3 but *in sets of* 4 *not in sets of* 5.

$$(0_1) = \{6, 3, 9, 12\},$$
$$(1) = \{1, 2, 4, 8\},$$
$$(2) = \{7, 14, 13, 11\}.$$

The table is then

$\times$	(0_1)	(1)	(2)
(0_1)	(0_1)	(0_1)	(0_1)
(1)	(0_1)	(1)	(2)
(2)	(0_1)	(2)	(1) .

(d) The product of any element of (0_1) and any element of (0_2) is zero.

(e) The sum of their identity elements is 1.

There is thus a very close analogy between Z_{15} organised in this way and the Fibonacci ring (mod 11), the chief difference being that the subfields are of different orders 3 and 5 whereas they are of the same order in the Fibonacci cases.

To complete the analogy it can be shown that the ideal consisting of (0_1) and zero organises the entire Fibonacci ring into cosets of 11 for multiplication. The first coset is obtained by adding 1 to each element of the ideal

$$(1) = \{1, 3f+3, 2f+6, \ldots, 10f+4\},$$

and the remaining cosets come from adding 2, 3, 4, ... to each element of the ideal. The quotient ring is the field Z_{11} but now the Fibonacci character is missing. The Fibonacci rule does not apply to these cosets.

145

Exercise

7.3 Investigate the sequence S_f using each of the moduli 13, 17, 19 in turn to reduce the terms.

(a) Decide in each case whether the modulus behaves like modulus 7 or modulus 11.

(b) If any of them appears to behave like modulus 7 confirm that the non-zero elements form a group for multiplication, indicating a normal subgroup and its cosets and naming its quotient group. Also find the reciprocal of $(f+3)$ or any other element of your choice of the form $af+b$ ($a \neq 0$ and $b \neq 0$).

(c) If any modulus appears to behave like modulus 11, determine the two smaller sets 0_1 and 0_2. Does each form a group for multiplication? If so find the identity element and arrange the group as a sequence of powers of one element.

7.10 Field or ring? Various criteria

In section 7.6, it was observed that the Fibonacci sequence seemed to produce two distinct patterns according to the modulus used.

Set I. Moduli leading to finite fields

Modulus	Number of cycles	Length of cycles
3	1	8
7	3	16
23	11	48
p	$(p-1)/2$	$2(p+1)$

Set II. Moduli leading to commutative rings

Moduli 11, 19, 31, 41, all lead to rings which have addition groups of the type $C_p \times C_p$ but the non-zero elements split into 3 multiplication groups, one with $(p-1)^2$ elements and two of the type C_{p-1}. How can it be foretold whether a modulus will lead to a field or a ring? And in the latter case how can the elements of the ideal be found?

One method is to look for zero divisors. If the property mentioned in section 7.8(v) holds good, there should be two elements $(f+a)$ and $(f+b)$, one from each ideal whose product is zero.

Hence
$$f^2+(a+b)f+ab=0,$$
$$(1-f)+(a+b)f+ab=0,$$
$$f(a+b-1)+ab+1=0.$$

146

Since f is irrational and a and b are integers, $a+b=1$ and $ab=-1$. This is equivalent to saying that a and b are roots of the congruency

$$x^2-x-1\equiv0$$

whose solutions if they exist are

$$x=(1\pm\sqrt5)/2.$$

This implies that the solutions exist if and only if 5 is a quadratic residue of the modulus p (see section 3.2). Taking two particular cases, 5 is not a quadratic residue if $p=7$, but if $p=11\sqrt5=4$ or 7, leading to $x=8$ or $x=4$ as solutions. The result is confirmed in table 7.3 where $(f+8)$ and $(f+4)$ appear as elements of K and L, M. Now $(f+8)$ can be used to generate the whole of K and $(f+4)$ leads rather less easily to L, M.

This method does not immediately reveal the identity elements and a better approach therefore is to search for these elements directly. If $cf+d$ is the identity element of K or L, M, then

$$(cf+d)^2=(cf+d).$$

This leads to two relations between c and d

$$c^2+d^2=d \tag{7.2}$$

and $\qquad\qquad\qquad c+c^2=2cd,$

i.e. $\qquad\qquad\qquad 1+c=2d$, since $c\neq0.$ $\qquad\qquad$ (7.3)

Eliminating c from (7.2) and (7.3)

$$5d^2-5d+1=0, \tag{7.4}$$

whence $\qquad\qquad\qquad d=(5\pm\sqrt5)/10.$

Once more the condition for solutions to exist is that 5 should be a quadratic residue of p. In the particular case when $p=11$, $\sqrt5=\pm4$,

hence $\qquad\qquad\qquad -d=5\pm4,$

$$d=2 \text{ or } 10,$$

whence $\qquad\qquad\qquad c=3 \text{ or } 8,$

so that identity elements must be $3f+2$ and $8f+10$ which can be confirmed in section 7.7.

Further, if d_1 and d_2 are solutions to equation (7.4) in the general case, then

$$d_1+d_2=1, \tag{7.5}$$

and since $\qquad\qquad\qquad c_1=2d_1-1,$

$$c_2=2d_2-1,$$

then $\qquad\qquad\qquad c_1+c_2=2(d_1-d_2)-2,$

$$c_1+c_2=0. \tag{7.6}$$

Equations (7.5) and (7.6) corroborate generally the observation made in section 7.8 that the sum of the identity elements is 1.

147

7.11 Condition for 5 to be a quadratic residue of p

By two different methods we have seen that the structure appears to be a ring if 5 is a quadratic residue (Q.R.) of p. This as it happens is an easy criterion to apply. First, there is no difficulty in deciding the inverse question whether p is a Q.R. of 5. Any number congruent to 1 or 4 (mod 5) is a Q.R. of 5 while numbers congruent to 2 or 3 are not. It follows that any prime number ending in digits 1 or 9 are Q.R.'s of 5 while those ending in 3 or 7 are not. There is a useful standard notation which helps to express these facts concisely:

$\binom{n}{q} = \quad 1$ means that n is a quadratic residue of the prime number q,

while

$\binom{n}{q} = -1$ means that n is a quadratic non-residue of q;

in the particular case above

$$\binom{p}{5} = \quad 1, \text{ if } p = 11, 19, 29, 31, \ldots,$$

$$\binom{p}{5} = -1 \text{ if } p = 7, 13, 17, 23, \ldots.$$

For the next step we can refer to the famous 'theorem of quadratic reciprocity' (proved in D. E. Littlewood, *A University Algebra* p.134, and other texts) which relates $\binom{p}{5}$ to $\binom{5}{p}$. The theorem states that if p and q are odd primes, then

$$\binom{p}{q} \times \binom{q}{p} = (-1)^{(p-1)(q-1)/4}.$$

If $q = 5$ this reduces to

$$\binom{p}{5} \times \binom{5}{p} = (-1)^{p-1} = 1, \text{ since } p \text{ is odd.}$$

Hence 5 is a Q.R. of p, if and only if p is a Q.R. of 5. This means that 5 is a Q.R. of p for values of p ending in 1 or 9 such as 11, 19, 29, $\ldots$ and not for numbers such as 7 and 13.

We have now arrived at a very simple criterion: a prime modulus produces Fibonacci rings for numbers ending in 1 or 9 and Fibonacci fields for numbers ending in 3 or 7.

148

7.12 Relation between p and the Fibonacci sequence

There is another method of approach which both discovers the sets of elements which form the ideals and also explains the origin of the two small Fibonacci sequences which alternate in one of the ideals. If we revert to modulus 11 for further study we see that the addition group contains C_{11} as a sub-group and the multiplication groups are all related to C_{10}. Hence the whole structure is closely related to the finite field Z_{11}. The non-zero elements of Z_{11} displayed as powers of 2 in figure 7.2 with one of its automorphisms in figure 7.3 show the connection with (0) and (0_1) in section 7.7.

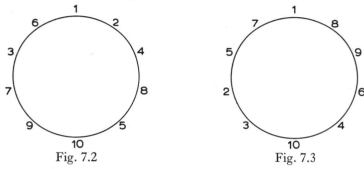

Fig. 7.2   Fig. 7.3

Sequences L and M come from alternate points in figure 7.2 and sequence K from consecutive points in figure 7.3. The technique may be used in reverse to find whether modulus 19 leads to a finite field or a commutative ring and in the latter case to find the elements in (0_1) and (O_2) which are the non-zero elements of the ideals.

The non-zero elements of Z_{19} may be arranged in powers of 2 and it is more convenient to use a rectangular rather than a circular dial:

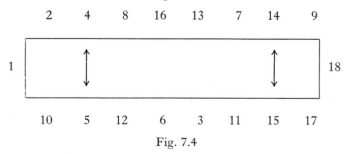

Fig. 7.4

Fibonacci sequences arise from this arrangement only if two consecutive numbers appear at equal distances from 1. In this case a pair 14, 15 are distant 7 from 1 and another pair 4, 5 distant 2 from 1.

149

As 7 is prime to 18, the sequence 14, 1, 15, . . . runs through all the 18 numbers while 4, 1, 5 recurs after 9 elements only. Between them they produce the two sets (0_1) and (0_2) which with zero form the two ideals of the ring. Starting each from its identity element I they are:

(0_1)	(0_2)
$2f+11$ (I)	$17f+ 9$ (I)
$11f+13$	$18f+14$
$13f+ 5$	$9f+ 7$
.	.
.	.
.	$4f+ 1$
$14f+ 1$	$2f+10$
$f+15$	$f+ 5$
.	.
.	.
.	.
$9f+ 2$	$11f+17$

Had we attempted to apply this technique with modulus 7 we should have found no suitable pair equidistant from 1.

Exercise

7.4 Find a sequence of powers of a residue class mod 31 which obeys the Fibonacci rule that two successive terms add to make the following term. Use your result to find some member of an ideal of the Fibonacci ring which is associated with modulus 31 and confirm that if this element is multiplied by any other element of the ring of your choice it produces an element of the ideal.

7.13   Modulus 5 and Modulus 29

Although we have so far mentioned two possible patterns for modulus p in connection with Fibonacci series, there are some interesting variations.

Modulus 5 might be expected to produce 2 cycles of 12 or else 4 cycles of 4 in one group and 2 separate groups of 4. However 5 is the only prime modulus for which $5=0$, so that the two identity elements indicated by equation 7.5 telescope into one and there is a single ideal consisting of the five elements

$$0, f+3, 3f+4, 4f+2, 2f+1.$$

The remaining 20 elements form a single group C_{20} for multiplication. The cosets of the ideal obtained by adding in turn 1, 2, 3, 4, to its elements are:

$$
\begin{array}{llll}
1, f+4, & 3f, & 4f+3, & 2f+2 \\
2, f, & 3f+1, & 4f+4, & 2f+3 \\
3, f+1, & 3f+2, & 4f, & 2f+4 \\
4, f+2, & 3f+3, & 4f+1, & 2f.
\end{array}
$$

These are precisely the same cosets as would be obtained by splitting the group C_{20} into cosets in the ordinary way, taking every fourth pair in the Fibonacci sequence

$$
\underline{0\ 1}\ 1\ 2\ \underline{3\ 0}\ 3\ 3\ \underline{1\ 4}\ 0\ 4\ \underline{4\ 3}\ 2\ 0\ \underline{2\ 2}\ 4\ 1.
$$

The pairs leading to the first coset which is a normal subgroup in C_{20} have been underlined and it is easy to check the other cosets. In this one case the Fibonacci method of organising the ring coincides with the standard method. The quotient ring is Z_5 and there is an exact parallel with the ring Z_{25}. The two systems are isomorphic.

Modulus 29 leads to a ring but the cycles in the main group are of length 14 instead of 28 as might have been expected from the precedent set by 11 (10-cycle) and 19 (18-cycle). Examples are:

$$
\begin{array}{l}
0, 1, 1, 2, 3, 5, 8, 13, 21, 5, 26, 2, 28, 1; \\
1, 4, 5, 9, 14, 23, 8, 2, 10, 12, 22, 5, 27, 3.
\end{array}
$$

There are corresponding modifications in the patterns for (0_1) and (0_2). The identity elements can be found using the method of section 7.10: they are $(8f+19)$ and $(21f+11)$. As before the automorphisms of Z_{29} are a useful tool for finding the elements of (0_1) and (0_2). They may be seen most conveniently when the 28 residues are arranged as powers of 8 and 10 respectively.

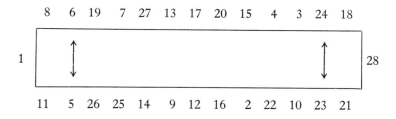

Fig. 7.5

151

The Fibonacci rule for (0_1) acts on alternate elements of the cyclic sequence in figure 7.5 and for (0_2) it operates on every fourth element in figure 7.6.

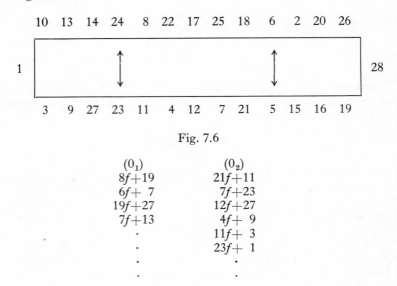

10 13 14 24 8 22 17 25 18 6 2 20 26

1

28

3   9   27 23 11 4 12 7 21 5 15 16 19

Fig. 7.6

(0_1)
$8f+19$
$6f+\ 7$
$19f+27$
$7f+13$
$\cdot$
$\cdot$
$\cdot$
$\cdot$

(0_2)
$21f+11$
$7f+23$
$12f+27$
$4f+\ 9$
$11f+\ 3$
$23f+\ 1$
$\cdot$

7.14   Reduction of S_f by a Composite Modulus

So far we have not examined any structure which might arise from reduction of the Fibonacci sequence S_f by a composite modulus. The exercise below contains some signposts for an investigation of what happens when 6 is used as a modulus.

Exercise

7.5   There are thirty-six different elements of the form $af+b$ where a and b are residue-classes mod 6. Let Q be this set of elements.

(a) Show that Q is a group for addition.

(b) Verify that under the Fibonacci rule the non-zero elements form three distinct recurring sequences, A, B, C, such that $A=1, f+1, f+2$, $2f+3, 3f+5, 5f+2, \ldots$ and B has more elements than C.

(c) Let B'=set B together with zero; let $C'=C$ with zero. Verify that B' and C' are normal subgroups of Q for addition and find their quotient groups.

(d) Verify that B' and C' are both fields for addition and multiplication and find their unity elements.

(e) Verify that Q is a ring for addition and multiplication and specify two ideals in Q.

(f) Find the quotient ring Q/B' and determine whether it is a field or a ring. Do the same for $Q/C.'$

(g) What is the kernel of the homomorphism which exists from the ring Q on to the quotient ring Q/B'?

(h) There is a homomorphism from the multiplicative structure of A and B combined onto the multiplicative structure of Z_4. Set out this homomorphism in full. What is its kernel?

8

POLYNOMIALS AND FINITE FIELDS

8.1 Further development of fields and congruencies

In section 7.1 a field was defined and examples were mentioned which included the finite field Z_p obtained by forming the residue classes for a prime modulus p. In this chapter we shall consider other finite fields obtained by combining Z_p with powers of x to form polynomial expressions in x: but first it is advisable to extend the ideas on congruency and on the solution of congruencies which were introduced in Chapter 1. Section 1.10 is particularly relevant.

8.2 Solutions to equations and congruencies

In this section congruency and equality must be carefully distinguished and the appropriate symbol will be used.

The equation $x^2-2x+3=0$ has no solution among real numbers; but if we choose to say that x is a residue class modulo 3 and to convert the equation to a congruency, then there are two solutions 0 and 2. That is to say, that any number in the set 0, $\pm$ 3, $\pm$ 6, . . . or in the set . . . , -4, $-1,2,5$, . . . satisfies the congruency $x^2-2x+3\equiv0$ which may now be more simply written as $x^2-2x\equiv0$ (mod 3). It is illuminating to look at this situation graphically. Let $y=x^2-2x+3$.

The graph may be read in two ways:

(a) *x, y real*. In this case the graph is continuous, and it is obvious that y lies wholly above the x axis and is never zero for any real value of x.

(b) *x a residue class modulo* 3. In this case x is restricted to the values 0, 1, and 2 or any integers congruent to 0, 1, or 2. Now $y\equiv0$ (mod 3) whenever the curve crosses one of the set of horizontal lines $y=0$, 3, 6, . . . However, not every crossing point is a solution of the congruency $x^2-2x+3\equiv0$ because, for example, when $y=9$, x does not take an integral value but is one of the irrational pair $1\pm\sqrt{7}$. The fact is that with x defined as an integer class mod 3, the graph of x^2-2x+3 consists merely of isolated integral points $(0, 3)$, $(1, 2)$, $(2, 3)$, $(3, 6)$. Two out of every three of these satisfy the congruency and have been ringed in figure 8.1. Other points are marked with a cross.

154

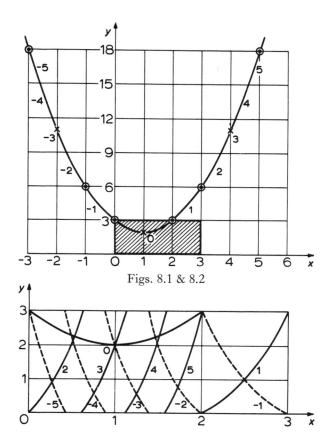

Figs. 8.1 & 8.2

8.3 A third aspect of the graph

It is possible to extend the idea of congruency modulo 3 to all real
numbers if we remove the restriction that x and y must be integers.
Then each real number is congruent to some real number x within the
range $0 \leq x < 3$, e.g. $4 \cdot 72 \equiv 1 \cdot 72$. With the same convention for y every
point in the Cartesian plane is congruent to some point in the rect-
angular space defined by $0 \leq x < 3$ and $0 \leq y < 3$. Figure 8.2 shows the
space enlarged and also displays several segments, numbered for
identification, of the discontinuous curve formed by transferring every
point on the original curve to its congruent image in the space. In other
words, one effect of the congruency is to pack the whole curve into a
single rectangle while preserving the separate identity of each segment.

It can now be seen that if x is defined as a real number there are an

155

infinite number of solutions to the congruency $x^2+2x\equiv 0$ (mod 3). The end points of every segment give possible solutions for x; and it is likely that an infinite number of other solutions would arise from considering other curves such as $4x^2-x-6=0$ which might be represented by the congruency. The usual condition that quadratic equations have at most two solutions no longer holds.

A merit of this interpretation is that it gives some meaning to dy/dx even when we are considering the isolated points of the graph interpreted as in section 8.2(b). For each specific curve represented by the congruency a continuum exists and a derivative exists at all points including the integral points. It seems not unreasonable to assign the same value for the derivative at these points whether the curve is regarded as a continuum or a chain of isolated points. As an example of the use of derivatives for integral congruencies consider

$$F(x)=x^2+x+1\equiv 0 \text{ (mod 3)}.$$

This has only one solution $x\equiv 1$ which may be accounted for algebraically by rewriting the congruency as $x^2-2x+1\equiv 0$. This implies that $(x-1)^2\equiv 0$ and suggests a pair of equal roots to the congruency. If we resort to calculus the condition for equal roots at $x\equiv 1$ would be that both $F(x)$ and $F'(x)$ should be zero at $x\equiv 1$. Since $F'(x)=2x+1$, which is congruent to zero at $x\equiv 1$, the condition is fulfilled and the use of calculus seems to be justified empirically in this case.

8.4 Polynomials over a field

This is a large and complex topic so a very informal treatment of the theory is given here—only just enough background to enable readers to embark on some of the simpler applications to finite groups.

A polynomial form familiar in elementary algebra is

$$ax^n+bx^{n-1} \ldots +k,$$

where n is a positive integer or zero, $a, b, c, \ldots k$ are integers and x may be an integer, a rational number, a real number or a trigonometrical function such as sine. To say that x is an indefinable or is indeterminate is to leave its nature wide open to many interpretations. It might be something more complicated such as a matrix.

We come now to 'polynomials over a field', a classification which is quite specific and implies that while x is indeterminate, and n a positive integer or zero, the coefficients $a, b, \ldots k$ must be elements of a field, possibly the rationals or reals, possibly the finite field formed by the residues to a prime modulus p. As an example of polynomials over a finite field we might take n as 2 and a, b, c from the residue classes mod 2, i.e. 0 and 1. There are exactly 8 such polynomials which are $x^2, x^2+x, x^2+1, x^2+x+1, x, x+1, 1, 0$.

The infinite set of polynomials over the rational numbers closely resembles integers in arithmetic and obeys similar rules. Both conform to all the requirements for a field except that of possessing an inverse for multiplication: there is no polynomial which will serve as a multiplier to convert x^2+3 into the multiplicative identity 1. For this reason they are classed as an *integral domain* (see definitions in Appendix 1).

Again, like integers, polynomials may be either composite, possessing factors, or prime. A prime polynomial is said to be irreducible and it is important to realise that a polynomial may be reducible over one field but not over another: x^2-3 is irreducible over the rationals but has factors $x-\sqrt{3}$, $x+\sqrt{3}$ over the reals.

In arithmetic if an integer a is divided by another integer d then the remainder r is less than the divisor d. In algebra the corresponding property is that the degree of the remaining polynomial is less than that of the divisor. Let us suppose that the polynomial x^4+3x^3+2x over the finite field of integers modulo 5 is to be divided by $2x^3+x$. We shall need the multiplication table of the field to help us.

$\times$	1	2	4	3
1	1	2	4	3
2	2	4	3	1
4	4	3	1	2
3	3	1	2	4

Since $3\times2=1$, the first term in the quotient is $3x$.

$$
\begin{array}{r|l}
& 3x+4 \\
\hline
2x^3+x & x^4+3x^3+2x \\
& x^4+3x^2 \\
\hline
& 3x^3+2x^2+2x \\
& 3x^3+4x \\
\hline
& 2x^2+3x
\end{array}
$$

The remainder $2x^2+3x$ is of less degree than the cubic divisor.

How many different remainders are possible after division by a cubic over this field? Since a quadratic has 3 terms and each of them may have any one of the coefficients 0,1,2,3,4, there are 5^3 possible remainders.

Consider now the much smaller set of polynomials of first degree or less when 2 is taken as the modulus of the field. There are only 2^2 possibilities: $x+1$, x, 1, 0. Their addition table is given below and it may easily be recognised as a Klein's group.

+	0	1	x	$x+1$
0	0	1	x	$x+1$
1	1	0	$x+1$	x
x	x	$x+1$	0	1
$x+1$	$x+1$	x	1	0

The multiplication table can also be written down but it does not look very promising at first sight.

$\times$	0	1	x	$x+1$
0	0	0	0	0
1	0	1	x	$x+1$
x	0	x	x^2	x^2+x
$x+1$	0	$x+1$	x^2+x	x^2+1

So far we have not committed ourselves to any value for x, but now, by exercising a deliberate choice, this table can be reduced to order. It is evident that x^2, x^2+x, and x^2+1 are reducible polynomials because they arose from multiplication of two linear factors. There is one other quadratic expression, x^2+x+1, which has no factors and so is an irreducible polynomial of the second degree, over the finite field, 0, 1, mod 2. If we choose to say that x satisfies the congruency $x^2+x+1\equiv0$, then each of the quadratic expressions in the table can be reduced in degree.

$$x^2\equiv x+1 \text{ (mod 2)},$$
$$x^2+x\equiv 1 \quad \text{(mod 2)},$$
$$x^2+1\equiv x \quad \text{(mod 2)}.$$

Now the table looks better:

$\times$	0	1	x	$x+1$
0	0	0	0	0
1	0	1	x	$x+1$
x	0	x	$x+1$	1
$x+1$	0	$x+1$	1	x

The non-zero elements 1, x, $x+1$ are seen to form a cyclic group of order 3 and the compact little set of four elements conforms to all the axioms of a field including the requirement that each non-zero element shall possess an inverse. The element x is the inverse of $x+1$ and 1 is self-inverse. The four elements form a field.

Referring to section 8.3 we can see that to satisfy the congruency $x^2+x+1\equiv0$, x might be given any one of an infinite number of values among real numbers. Among these are the two Fibonacci numbers f and $f+1$ which we already know satisfy the equations $x^2+x-1=0$ and $x^2-x-1=0$ respectively and so satisfy $x^2+x+1\equiv0$ (mod 2). In detail,

if
$$x=f+1,$$
then
$$x^2+x+1=(f+1)^2+(f+1)+1$$
$$=(f+2)+(f+1)+1$$
$$\equiv0\,(\text{mod } 2).$$

But x might also be a real root of $3x^2-5x-7=0$ or any of a host of other equations which reduce to $x^2+x+1\equiv0$ (mod 2).

Let us now designate by R a particular solution of the congruency. Then $R^2+R+1\equiv0$ or more conveniently $R^2\equiv R+1$. This last form is useful for reducing any power of R to a linear form:

$$R^3=R\times R^2$$
$$\equiv R(R+1)$$
$$\equiv R^2+R$$
$$\equiv(R+1)+R$$
$$\equiv1,$$
$$R^4\equiv R,$$
$$R^5\equiv R+1,$$
$$\cdots\cdots$$

It can be seen at once that the powers of R form a recurrent cycle of order 3 and it is generally true, though we shall not prove it here, that an irreducible congruency of degree p where p is a prime number, over

159

a finite field of 0, 1, . . . m (m also prime) generates m^p-1 different functions of R where R satisfies the congruency. Each function of R may be obtained as a distinct power of R using the method just given. For example, if $p=5$ and $m=2$, and R is a solution of $x^5 \equiv x^2+1$, then $R^5 \equiv R^2+1$, and it follows that

$$
\begin{aligned}
R^6 &= R\,(R^2+1) \\
    &= R^3+R, \\
R^7 &= R^4+R^2, \\
R^8 &= R^3+R^2+1, \\
R^9 &= R^4+R^3+R \\
    &= (R^2+1)+R^4+R^2 \\
    &= R^4+1,
\end{aligned}
$$

and so on through the 2^5-1 possible functions of R. With zero added these same functions become the 2^5 residues to any irreducible fifth degree polynomial in R over the same field 0, 1 (mod 2). Each residue is of the form $aR^4+bR^3+cR^2+dR+e$. There are 2^5 in all because each of $a, b, \ldots e$ may be either 0 or 1.

The thirty-two elements form a finite algebraic field satisfying all requirements for a field. It is well established that fields may be formed in this way by combining an arithmetic field and a solution R to an irreducible congruency over the field. Such a field is called a Galois field in honour of the young French mathematician who left behind him important ideas on groups and fields before being so tragically killed in a duel. The particular field we have just been discussing is called G.F. 2^5 (since $m=2$ and $p=5$).

The rest of this chapter will be concerned with ideas arising from geometrical interpretations of such fields.

8.5   Geometrical interpretation of G.F.2^2

The field G.F.2^2 has already been described in some detail at the beginning of the previous section using the symbol x instead of R, which we shall now adopt permanently to remind ourselves that it may be regarded both as a residue to an irreducible polynomial and also a root of an irreducible congruency. Its behaviour as an additive group may be illustrated geometrically by assigning each of the four elements 0, 1, R, and $R+1$ to a corner of a square:

Fig. 8.3

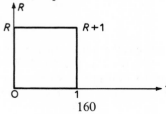

If the side of the square has any convenient length for unit and if R and 1 are associated respectively with two directions at right angles then starting at 0 and moving 1 unit in the R direction leads to the point R,

i.e.
$$0+R=R;$$

in the same spirit
$$0+1=1$$
$$1+R=R+1$$
$$(R+1)+R=2R+1$$
$$\equiv 1.$$

To continue this process indefinitely leads to the construction of a lattice in which each point differs from its neighbour by a single step either in the 'R' direction or the '1' direction. Also by virtue of the congruency the name of every point in the lattice is either $R+1$, R, 1, or 0. By mapping the whole plane in this way it can be seen that it consists of 4 distinct lattices of scale 2 units.

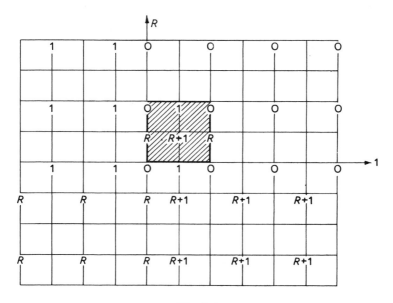

Fig. 8.4

A sample square (2 units) in the diagram has been shaded and out-lined in bold and the names of all points on the 1-unit lattice inserted. In other squares only one point has been named, different for each quadrant so that the congruencies of the four resulting lattices may be easily distinguished.

Consider now the addition $(R+1)+R$. Algebraically:

$$(R+1)+R=2R+1$$
$$\equiv 1.$$

Geometrically: begin at any $(R+1)$ point and proceed one step in either R direction; then you must come to a point on the lattice of 1's. Note that just as the lattice consists of isolated points so movement is restricted to two directions only. Direct motion from 1 to R across the diagonal of a square is meaningless in this representation.

8.6 Geometrical representation of G.F.2^3

(a) *Addition*

A similar three-dimensional representation gives a good picture of the addition for the 8 elements of the field G.F. 2^3. The three axes are associated respectively with 1, R, and R^2 (Sawyer, A Concrete Approach to Abstract Algebra, p.128).

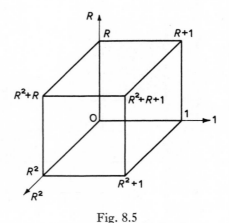

Fig. 8.5

This addition group has been discussed algebraically in section 3.11.

162

Is it possible to use an alternative method, associating 1, R, and R^2 with the axes for triaxial co-ordinates in a plane?

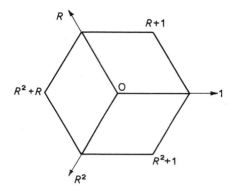

Fig. 8.6

Up to a point the representation is successful. The six elements 1, $R+1$, R, R^2+R, R^2, R^2+1, appear at the corners of a regular hexagon but the snag comes over 0 and R^2+R+1 which coincide at the centre of the hexagon. The reason for this can be seen by comparing the two diagrams: the plane representation is really one aspect of the cubic one, looking along the diagonal joining 0 to R^2+R+1.

Another serious defect in the representation is that it makes the point R appear to be congruent to R^2+1, differing from it by two units in a permissible direction. Reference to the first diagram shows that the only permissible routes from R to R^2+1 are by three separate steps

such as $R \longrightarrow R+1$, $R+1 \longrightarrow 1$, $1 \longrightarrow R^2+1$,

i.e. $$R+1-R+R^2 \equiv R^2+1.$$

The route across the hexagon $R \longrightarrow 0$, $0 \longrightarrow R^2+1$, conceals a forbidden step $0 \longrightarrow (R^2+R+1)$ across the diagonal of the cube in the first representation. Although the representation is defective it has been inserted as an introduction to similar representations in G.F.2[5] and G.F.2[7].

It can easily be confirmed by compiling an addition table that the addition group is of type $C_2 \times C_2 \times C_2$ described in section 3.11.

163

(b) *Multiplication*

For multiplication the seven non-zero elements form a cyclic group, best represented on a clock. Remembering that R in the field must be the root of an irreducible cubic congruency and noting that the only possible ones are $x^3 \equiv x+1$ and $x^3 \equiv x^2+1$, we select the former as being a little more convenient to work with:

$$R^3 \equiv R+1,$$
$$R^4 \equiv R(R+1) = R^2+R.$$

Continuing in this way, $R^5 \equiv R^2+R+1,$
$$R^6 \equiv R^2+1,$$
$$R^7 \equiv 1.$$

Hence the 7 elements may be displayed as powers of R:

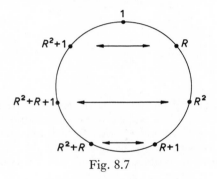

Fig. 8.7

Reciprocal elements have been connected by arrows. The product is 1 for each pair.

8.7 Geometrical representation of G.F.2^5

Addition. As there is now little danger of ambiguity the congruency sign will be dropped in favour of the more familiar equals sign.

The first representation in section 8.6 fails for higher dimensions but the second one, defective though it is, may be extended to G.F.2^5. It successfully displays 30 of the 32 elements of the field in three concentric circles but the remaining two elements 0 and $R^4+R^3+R^2+R+1$ coincide at the centre of the circles.

164

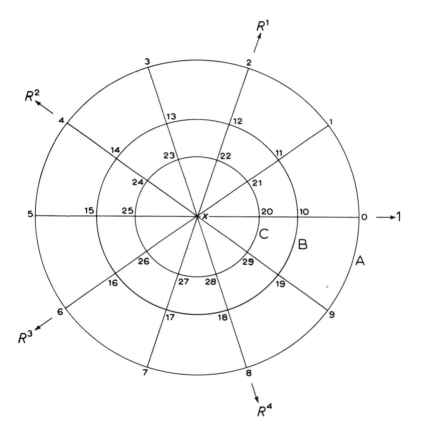

Fig. 8.8

The figure shows five axes at angles of 72° associated respectively with
1, R, R^2, R^3 and R^4. The points have been numbered and the key to

their names is given below.

$$
B \begin{cases}
10 & 1 \\
11 & 1 +R +R^2+R^4 \\
12 & R \\
13 & 1 +R +R^2+R^3 \\
14 & R^2 \\
15 & R +R^2+R^3+R^4 \\
16 & R^3 \\
17 & 1 +R^2+R^3+R^4 \\
18 & R^4 \\
19 & 1 +R +R^3+R^4
\end{cases}
\qquad
C \begin{cases}
20 & R +R^4 \\
21 & 1 +R +R^3 \\
22 & 1 +R^2 \\
23 & R +R^2+R^4 \\
24 & R +R^3 \\
25 & 1 +R +R^3 \\
26 & R^2+R^4 \\
27 & R +R^3+R^4 \\
28 & 1 +R^3 \\
29 & 1 +R +R^4
\end{cases}
$$

$$
A \begin{cases}
0 & 1 +R +R^4 \\
1 & 1 +R \\
2 & 1 +R +R^2 \\
3 & R +R^2 \\
4 & R +R^2+R^3 \\
5 & R^2+R^3 \\
6 & R^2+R^3+R^4 \\
7 & R^3+R^4 \\
8 & 1 +R^3+R^4 \\
9 & 1 +R^4 \quad,
\end{cases}
$$

$$x \quad 0 \quad 1 +R +R^2+R^3+R^4.$$

The intermediate circle B is taken to have unit radius. To find the point $R+R^2+R^4$ one should begin at X and obtain the product of successive unit displacements in directions R, R^2, and R^4. The final result is at point 23.

There are two interesting features of figure 8.8.:

(i) If the intermediate circle B has unit radius the two other circles A and C have as radii the two Fibonacci constants $f+1$ and f. This follows at once from the pentagonal nature of the diagram. The characteristic triangles 36°, 72°, 72° and 108°, 36°, and 36° are obvious in figure 8.9.

(ii) The coincidence of the two centre points 0 and $1+R+R^2+R^3+R^4$ suggests that the figure is really a projection of a three-dimensional pattern of points, and this turns out to be quite a fruitful idea. Good diagrams of both the icosahedron and the dodecahedron can be obtained by making suitable connections among the 32 points, which for brevity we will now call pattern F.

The projection of the icosahedron consists of the ten points of circle A with the point X which is really the projection of two distinct points

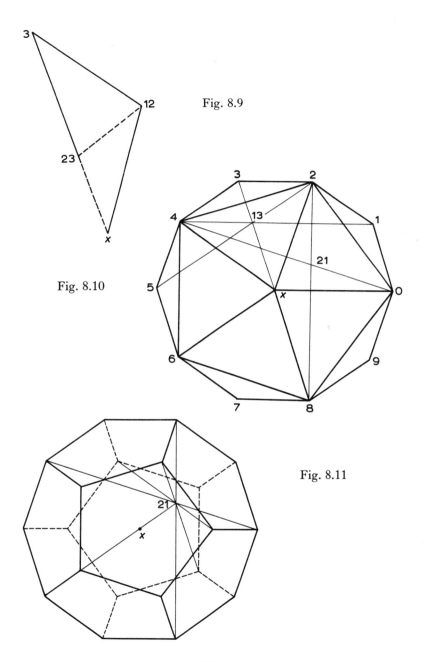

Fig. 8.9

Fig. 8.10

Fig. 8.11

167

X' and X in space (X nearer to the reader), each lying on an axis perpendicular to the plane of the paper. The remaining 20 points of pattern F on circles B and C may then be identified as projections of the points where the internal diagonals of the solid intersect, one located below each of the 20 faces. For example, point 21 lies below the triangle X, 0, 2 on the diagonals 0–4, 2–8, X–1; and point 13 lies below 2, 3, 4 on diagonals 2–5, 3–x' and 4–1. Another way of describing these 20 points would be as the diagonal points of various plane pentagons formed by the vertices of the icosahedron such as 0, 2, 4, 6, 8 or X, 2, 1, 9, 8. From considerations of symmetry the 20 points must, of course, form a dodecahedron.

In Fig. 8.11 the situation is reversed. The outer two circles of pattern F give the projection of a dodecahedron and the inner circle, with the two points X' and X on a central axis, that of an icosahedron formed by the intersections of its diagonals. Thus the same plane configuration pattern F may be regarded as the projection of two distinct configurations in space, 8.10 corresponding to a dodecahedron within an icosahedron and 8.11 to a icosahedron within a dodecahedron. Again each of these space-configurations may be interpreted in two ways, the two for 8.11 being (i) as described above, an outer dodecahedron with inner diagonal points; (ii) a great stellated dodecahedron made by producing the edges of an inner icosahedron (Fig. 8.12).

This is classified as a regular solid each of whose faces is a star pentagon, three such faces meeting at each of the 20 true vertices. Points such as X and 21 where the edges intersect at intermediate points, not end points, are called false vertices. There are three other regular solids of this kind whose faces are respectively another star pentagon, a regular pentagon, and an equilateral triangle, each lying partly within the solid and all of them may be displayed quite clearly by appropriate use of pattern F. The names are 'great icosahedron', 'great dodecahedron', and 'small stellated dodecahedron'.

8.8   G.F.2^7

It is interesting to speculate beforehand whether G.F.2^7 can be represented in the same way for addition as G.F.2^5. There are 128 elements instead of 32 and if the special elements 0 and $1+R+R^2+R^3+R^4+R^5+R^6$ are omitted there remain 126. Resolving 126 into 14×9 it seems likely that the points of the field will be found on nine concentric circles along rays from the centre as for the pentagonal distribution. Investigation shows that this is not far from the truth. There are in fact 8 concentric circles and of these one presents features different from the rest: it contains 28 instead of 14 points and these do not lie on the central rays. That might be the end of the investigation

Fig. 8.12

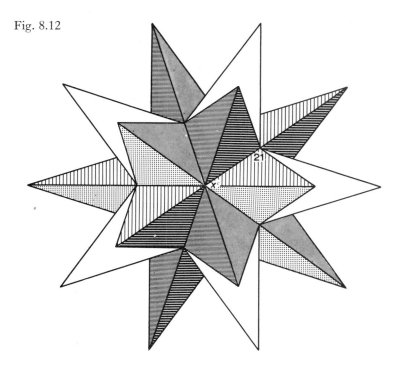

but there are some interesting relationships between the radii of the circles. For the pentagon of G.F.2^5 the radii of the three circles were f, 1, $f+1$ f being the Fibonacci constant $(\sqrt{5}-1)/2)$. Also f is the length of the side of a regular decagon inscribed in the unit circle, so that $f=2\sin(\pi/10)$: if we take h to be the length $2\sin(\pi/14)$ of the side of a regular 14-gon inscribed in a unit circle then the radii of the 8 circles in order of size are h, $1-h$, $h/1-h$, 1, $(1-h)/h$, $\sqrt{2}$ (this is the special circle with 28 points on it), $1/(1-h)$ and $1/h$. The expressions in h are strongly reminiscent of the t-expressions in Chapter 2:

$$t,\ 1-t,\ \frac{t}{t-1},\ \frac{t-1}{t},\ \frac{1}{1-t},\ \frac{1}{t}.$$

One begins to suspect the presence of the group D_3.

If so there is an analogy with the association of 1, f, $f+1$, $-f$ with Klein's four-group in its representation 1, t, $1/t$, $-t$, combined by substitution. See section 4.7.

The simplest way of establishing the connections between the radii is to consider the six triangles below which occur in Fig. 8.16. Angles are as marked and their sides can be calculated using similar triangles; $\theta = \pi/7$; L, M, N have unit sides; L', M', N' have unit bases; the base of L is h by definition of h; it happens that L, M, N fit neatly together to make the larger triangle L'.

Fig. 8.13

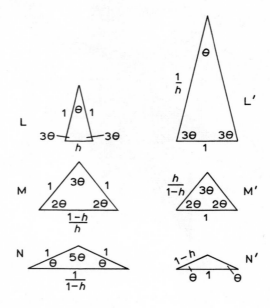

From similar triangles PSR, PQR,

since SR$=h$, $\qquad\qquad$ PQ$=\dfrac{1}{h}$, $\qquad\qquad$ Fig. 8.14

hence $\qquad\qquad$ PT$=\dfrac{1}{h}-1=\dfrac{1-h}{h}.$

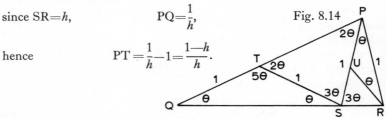

Draw RU making an angle θ with PR, so that SRU$=2\theta$.
Then SU$=$SR$=h$, making PU$=1-h$.
This identifies triangle PUR with N'.

Remaining lengths are obvious from comparing similar triangles in Fig. 8.13.

8.9 Diagram of G.F.2[7]

From the results of the last section it is easy to show the relationship between the radii of the 8 circles in Fig. 8.16. Eight typical points are shown in Fig. 8.15 below. X is on the unit circle, Y on the special circle, points A to F are at distances determined by h.

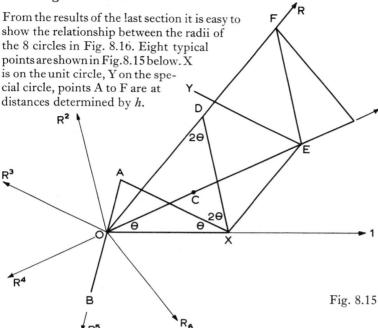

Fig. 8.15

They are listed in the order in which they can be derived.

Point	Distance from 0	Name	Comment
X	1	1	
A	h	$1+R^3$	triangle L
B	$1-h$	$1+R^3+R^5$	$OB=1-OA=1-h$
F	$\dfrac{1}{h}$	$1+R+R^2$	triangle L'
D	$\dfrac{1-h}{h}$	$1+R^2$	triangle M
E	$\dfrac{1}{1-h}$	$1+R$	triangle N
C	$\dfrac{h}{1-h}$	$1+R+R^4$	$OC=OE-1=\dfrac{1}{1-h}-1$ $=\dfrac{h}{1-h}$
Y		$1+R+R^3$	See 8.11

171

The relative positions of X,Y,A,B,C,D,E,F can be seen in Fig. 8.16.

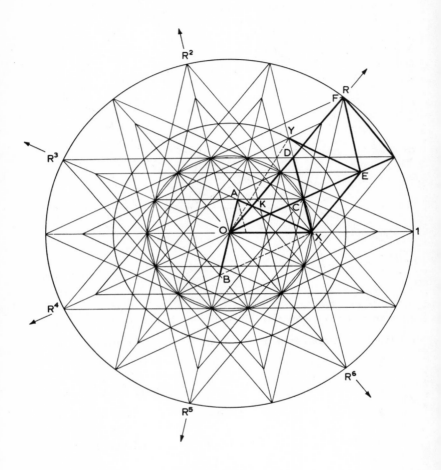

Fig. 8.16

8.10 Cubic equation for h

To find the radius of the special circle, having Y as a typical point on it, attention must be focused on h. It can be shown to satisfy a cubic equation.

Comparing triangles OKX and BOX in Fig. 8.16

$$\frac{OB}{OX} = \frac{OK}{KX},$$

$$(AK = h^2)$$

$$\frac{1-h}{1} = \frac{h}{1-h^2},$$

leading to $h^3 - h^2 - 2h + 1 = 0 \ldots$ (8.1)

We noted in section 8.8 that the radii of the six circles were akin to the t-expressions in Chapter 2. We now have an opportunity to establish firmly that h is a special value of t, because of the natural symmetry existing between the roots of equation (8.1). Not only h but also $1 - 1/h$ and $1/(1-h)$ satisfy the cubic:

$$\left(\frac{1}{1-h}\right)^3 - \left(\frac{1}{1-h}\right)^2 - 2\left(\frac{1}{1-h}\right) + 1$$

$$= \frac{1 - (1-h) - 2(1-h)^2 + (1-h)^3}{(1-h)^3}$$

$$= \frac{-h^3 + h^2 + 2h - 1}{(1-h)^3}$$

$$= 0, \text{ from equation (8.1).}$$

A similar proof shows that $(1 - 1/h)$ also satisfies the cubic. The remaining expressions in h, namely $(1-h)$, $1/h$ and $h/(h-1)$ satisfy the reciprocal cubic $h^3 - 2h^2 - h + 1 = 0$.

8.11 Radius of the circle with 28 points in Fig. 8.16

OY may now be determined from triangle OYE in Fig. 8.16.

$$OY^2 = 1 + OE^2 - 2 \cdot OE \cos 2\theta$$

$$= 1 + \frac{1}{(1-h)^2} - \frac{1}{1-h} \cdot \frac{1-h}{h} \quad (2\cos 2\theta = \frac{1-h}{h} \text{ in triangle M})$$

$$= 1 + \frac{1}{(1-h)^2} - \frac{1}{h}$$

$$= 2 + \left\{ \frac{1}{(1-h)^2} - \frac{1}{h} - 1 \right\}$$

$$= 2 + \left\{ \frac{-h^3 + h^2 + 2h - 1}{h(1-h)^2} \right\}$$

$$= 2 + 0, \text{ from equation (8.1),}$$

hence $OY = \sqrt{2}$.

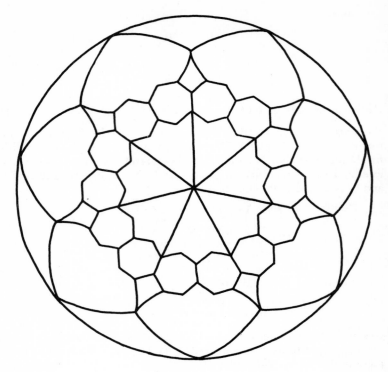

Fig. 8.17

8.12   Heptagonal symmetry

It seems certain that h is a constant associated with the heptagon as the Fibonacci constant belongs to the pentagon. Throughout the ages the numbers f, 1, and $f+1$ have been combined by artists and architects into many beautiful forms. Possibly there are other equally harmonious results to be obtained from heptagonal symmetry using h and its associated lengths in combination with 1 and $\sqrt{2}$. Fig. 8.17 shows a rudimentary attempt to make a conventional design from points of Fig. 8.16. It is for the reader to decide whether it possesses any natural symmetry derived from its association with the dihedral group D_3. Another possibility is that rectangular blocks with sides proportional to h, $(1-h)/h$, $1/(1-h)$ and $1-h$, $1/h$, $h/(h-1)$ may have harmonious properties in combination with each other and with heptagonal designs based on these lengths.

9

MAPPINGS, PERMUTATION GROUPS, AND GROUPS OF AUTOMORPHISMS

9.1 Mappings

We have already used the term 'mapping' in several places without defining it precisely. It is appropriate now to formalise the idea so that we may go on to study permutations which form a particular class of mapping and also to explore further the idea of automorphism.

A mapping is much the same as a transformation, the difference being rather one of context than meaning. The word 'transformation' is likely to occur in a geometrical context while 'mapping' is part of set language. We shall use the symbol t for a mapping in our definition to remind us of this connection.

A mapping t from a set A to a set B assigns to each member a of A a unique member b of B. Then b is said to be the image of a under the mapping.

Fig. 9.1

There is a difference between 'mapping into' and 'mapping onto'. In Fig. 9.1 the mapping is from A into B. If it had been onto, every b would have an arrow leading to it. In a crowded cloakroom with at least one coat on every peg the set of coats C is mapped onto the set of pegs P. The working of the attendant's mind or perhaps some pre-arranged system determines the mapping. If tickets are issued in pairs, one ticket for each coat and the duplicate for its owner, then the ticket system provides a different mapping from the set of owners to the set of coats. This new mapping is both one-one and onto. Again if we visualise the cloakroom as belonging to a theatre, it would be unlikely that every member of the audience would deposit a coat there. This means that yet another mapping might be specified from the set of coats to the set of people sitting in the auditorium; this mapping would be one-one and into.

A mapping always conveys a sense of direction, *from* A *to* B, and so is often symbolised by an arrow as in Fig. 9.1. If the mapping is both one-one and onto there is also an inverse mapping from B onto A since every member of B has a unique image in A. The arrow makes it natural to speak of a mapping as sending a in A to its image b in B. This is a useful mode of thought when we wish to combine two mappings. The rule of combination is succession. A mapping t_1 from A to B followed by a mapping t_2 from B to C produces a third mapping which we shall call $t_1 t_2$ and this is from A to C.

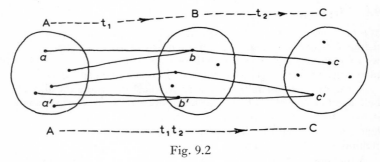

Fig. 9.2

In Fig. 9.2 it may be seen that the combined mapping $t_1 t_2$ sends a to c via b while a' goes to c' via b'. Each element in A has a unique image in C, either c or c'. We have established the fact that mappings are closed under succession.

In a mapping from A to B the set B need not be distinct from A, and it frequently happens that B either overlaps A, or is a subset of A or is identical with A.

Fig. 9.3

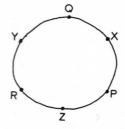

In Fig. 9.3 a rotation of 60° anticlockwise is a one-one mapping from the set A={P, Q, R} onto the set B={X, Y, Z} such that P——→X, Q——→Y, R——→Z but a rotation of 120° is a one-one mapping from A onto A such that P——→Q, Q——→R, R——→P.

9.2 Associative property of mappings

The rule of combination for mappings is associative. Consider a succession of three mappings, t_1 from A to B, t_2 from B to C and t_3 from C to D, as illustrated in Fig. 9.4.

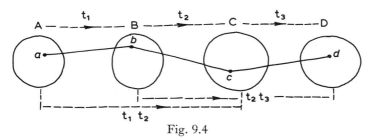

Fig. 9.4

To show that $(t_1 t_2)t_3 = t_1(t_2 t_3)$ we need only consider the fate of a typical element a in A.

Let $\qquad$ t_1 send a to b in B,
$\qquad\qquad\qquad$ t_2 send b to c in C,
$\qquad\qquad\qquad$ t_3 send c to d in D;

then $(t_1 t_2)t_3$ sends a to c and then from c to d,
i.e. $\qquad\qquad\qquad$ sends a to d in D;

and $t_1(t_2 t_3)$ sends a to b and then from b to d in D
i.e. $\qquad\qquad\qquad$ sends a to d in D.

Hence $(t_1 t_2)t_3 = t_1(t_2 t_3)$ for a and similarly for all other a's in A.

9.3 Permutations

A mapping of a finite set onto itself is necessarily one-one, otherwise it would be a mapping into itself. Another name for this very important kind of mapping is a permutation and a separate notation has been developed for handling permutations.

Consider a set of four playing cards, say Ace, King, Queen, Jack.

There are 24 possible arrangements of them and an operation which changes the order of an arrangement is called a permutation. Here are three arrangements:

1. A K Q J
2. K A J Q
3. Q A K J

In changing 1 into 2 Ace is replaced by King, King by Ace, Queen by Jack, and Jack by Queen. This can be written shortly (AK) (QJ), meaning that each of the two pairs is transposed. From 2 to 3 the permutation is King⟶Queen, Queen⟶Jack, Jack⟶King, while the Ace is unaltered. This is written (K Q J). From 1 to 3 the permutation is (A Q K) so that we may say that (K Q J) following on (AK) (QJ) is equivalent to (A Q K). This statement is true whatever initial arrangement of the four cards is used to demonstrate it. For example, if the cards are arranged J A Q K then:

(AK) (QJ) acting on J A Q K produces Q K J A,
(KQJ) acting on Q K J A produces J Q K A,
also   (A Q K) acting on J A Q K produces J Q K A.

Each permutation has its inverse, a permutation which restores the original position.

{(AK) (QJ)} is its own inverse,
but $(KQJ)^{-1}=(KJQ)$
and $(AKJQ)^{-1}=AQJK.$

Successive applications of a single permutation recur to the original arrangement after two, three, or four steps:

$(AK)^2=I$ and $\{(AK)(QJ)\}^2=I,$
$(AKQ)^3=I$ and $(AKQJ)^4=I.$

(AK) is a permutation of order 2. It is a transposition involving two letters only. (AKQ) is evidently a permutation of order 3; notwithstanding it is said to be an *even* permutation because it can be resolved into an even number of transpositions. To take an example, arrange the cards A K Q J: the permutation (AKQ) produces K Q A J and this result could also be produced by two transpositions, first (AK) and then (AQ). To prove that the number of transpositions required to transform a given arrangement A into another B is always even (or always odd as the case may be), whatever the sequence of transpositions used, is laborious and the fact which is well-established will be assumed here. The reader may like to verify that this is so in some particular instance.

A pack of cards is useful for verifying these basic facts since they can be arranged in an order easily recognisable, such as A K Q J using one suit, and the effect of permutation can be studied by making a rearrangement in another suit; but of course any symbols will serve and for paperwork it is a common practice to use numerals:

$$(1\ 3\ 4\ 2) \text{ acting on } 1\ 2\ 3\ 4 \text{ leads to } 3\ 1\ 4\ 2.$$

We see that a permutation which converts $1\ 2\ 3\ 4$ into $3\ 1\ 4\ 2$ may be expressed in a number of ways. It may be written $\begin{pmatrix} 1\ 2\ 3\ 4 \\ 3\ 1\ 4\ 2 \end{pmatrix}$ which sets out the mapping in detail in vertical pairs $1 \longrightarrow 3$, $2 \longrightarrow 1$, $3 \longrightarrow 4$, $4 \longrightarrow 2$. If it suits us to do so we may rearrange the order of the vertical pairs without affecting the mapping specified.

$$\begin{pmatrix} 1\ 2\ 3\ 4 \\ 3\ 1\ 4\ 2 \end{pmatrix} \text{ is exactly the same as } \begin{pmatrix} 3\ 1\ 2\ 4 \\ 4\ 3\ 1\ 2 \end{pmatrix}.$$

Finally it may be written more compactly as $(1\ 3\ 4\ 2)$ and read mentally as $1 \longrightarrow 3$, $3 \longrightarrow 4$, $4 \longrightarrow 2$, $2 \longrightarrow 1$.

Here is another example using 8 symbols:

$$\begin{pmatrix} 1\ 2\ 3\ 4\ 5\ 6\ 7\ 8 \\ 3\ 5\ 4\ 1\ 2\ 7\ 6\ 8 \end{pmatrix} = (1\ 3\ 4)(2\ 5)(6\ 7).$$

9.4 Combining permutations

Suppose we want to combine two permutations on 8 symbols.

For example, let
$$p_1 = (1\ 3\ 4)(2\ 5)(6\ 7),$$
$$p_2 = (1\ 4\ 5\ 8).$$

We want to find the product $p_1 p_2$ by which we mean the permutation which is equivalent in effect to that of p_1 followed by p_2. There are various methods open to us according to the skill attained in handling permutations.

(a) We can set out the mappings side by side in detail:

$$\begin{pmatrix} 1\ 2\ 3\ 4\ 5\ 6\ 7\ 8 \\ 3\ 5\ 4\ 1\ 2\ 7\ 6\ 8 \end{pmatrix} \begin{pmatrix} 1\ 2\ 3\ 4\ 5\ 6\ 7\ 8 \\ 4\ 2\ 3\ 5\ 8\ 6\ 7\ 1 \end{pmatrix} \qquad (9.1)$$

and then rearrange the pairs in the second bracket so that the sequence in the top row is identical with the sequence in the bottom row of the other bracket:

$$\begin{pmatrix} 1\ 2\ 3\ 4\ 5\ 6\ 7\ 8 \\ 3\ 5\ 4\ 1\ 2\ 7\ 6\ 8 \end{pmatrix} \begin{pmatrix} 3\ 5\ 4\ 1\ 2\ 7\ 6\ 8 \\ 3\ 8\ 5\ 4\ 2\ 7\ 6\ 1 \end{pmatrix} \qquad (9.2)$$

The resulting permutation is then quite plainly

$$\begin{pmatrix} 1\ 2\ 3\ 4\ 5\ 6\ 7\ 8 \\ 3\ 8\ 5\ 4\ 2\ 7\ 6\ 1 \end{pmatrix} \qquad (9.3)$$

obtained by missing out the connecting link $3\ 5\ 4\ 1\ 2\ 7\ 6\ 8$ and this can be expressed in cycles as

$$(1\ 3\ 5\ 2\ 8)\,(6\ 7). \qquad (9.4)$$

179

(b) We could omit the rearrangement and go straight to step (9.3) or even step (9.4).

(c) We could go straight from the given permutations

$$p_1 = (1\ 3\ 4)(2\ 5)(6\ 7),$$
$$p_2 = (1\ 4\ 5\ 8),$$

to the combined permutation $p_1 p_2$ by mental steps such as the following:
Begin with 1:

under p_1	$1 \longrightarrow 3$,
under p_2	$3 \longrightarrow 3$,
hence under $p_1 p_2$	$1 \longrightarrow 3$;
under p_1	$3 \longrightarrow 4$,
under p_2	$4 \longrightarrow 5$,
hence under $p_1 p_2$	$3 \longrightarrow 5$;

and so on, building up the first cycle (1 3 5 . . .) and then continuing with another cycle.

It is worth remarking in this example that the cycle (6 7) in p_1 goes through unchanged into $(p_1 p_2)$ since neither 6 nor 7 appear in p_2.

Exercises

9.1   Find $p_2 p_1$ and $p_1{}^2$.
9.2   Find $p_1{}^{-1}$, $p_2{}^{-1}$ and $p_1{}^{-1} p_2 p_1$.

We have just considered a particular example of a general truth that the product of any two permutations on n symbols is again a permutation on the n symbols. More shortly permutations of any arrangement of symbols are closed under succession. An adaptation of Fig. 9.2 for a set A consisting of four elements makes it plain why this is so.

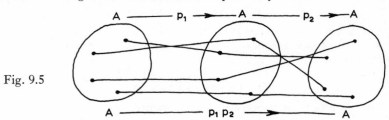

Fig. 9.5

A permutation of an arrangement of A is a one-one mapping from A onto itself. Since p_1 and p_2 are both one-one and onto, it follows that $p_1 p_2$ must also be one-one and onto. This has nothing to do with the number of elements, four, so the statement is generally true for a set of n symbols.

180

9.5 Symmetric groups

In the theory of groups a very important part is played by the *symmetric group of degree n* known shortly as S_n. Its elements consist of all possible permutations on n symbols and as before the rule of combination is succession. Readers are no doubt familiar with the fact that there are $n!$ different arrangements of n symbols and therefore also $n!$ different ways of permuting any given arrangement. Hence the order of the group S_n is $n!$ It can easily be verified that the permutations form a group under succession.

(a) Closure—established in section 9.4.

(b) Associative—established for mappings in 9.2.

(c) The identity permutation is $\begin{pmatrix} 1 \, 2 \ldots n \\ 1 \, 2 \ldots n \end{pmatrix}$

(d) Each permutation has an inverse found by reversing the order within each of its cycles:

$$[(1 \ 3 \ 4 \ 7)(2 \ 5 \ 8)^{-1} = (1 \ 7 \ 4 \ 3)(2 \ 8 \ 5)].$$

The product of these two is evidently the identity permutation.

We met the group S_3 in section 2.12. It is isomorphic with D_3 and its table is very familiar.

Table 9.1

second permutation

	*	1	2	3	4	5	6
	1	1	2	3	4	5	6
	2	2	3	1	5	6	4
first	3	3	1	2	6	4	5
permutation	4	4	6	5	1	3	2
	5	5	4	6	2	1	3
	6	6	5	4	3	2	1

If we regard this table as belonging to S_3, the group of all possible permutations on three symbols a, b, c then one way of assigning the

181

symbols 1–6 is as follows:

$$1 = \text{identity permutation} \quad 4 = (ab)$$
$$2 = (abc) \qquad\qquad\qquad 5 = 2 \ast 4 = (bc)$$
$$3 = 2^{-1} = (acb) \qquad\qquad 6 = 3 \ast 4 = (ca)$$

It is possible however to regard the same table as the source of a group of permutations on the *six* symbols 1, 2, 3, 4, 5, 6. The group concerned is a subgroup of S_6 and is isomorphic with S_3. It is formed as follows.

Since every column of the table is a permutation of the first column we can define 6 permutations of degree 6:

$$p_1 = \text{the identity permutation, from column 1 to column 1,}$$
$$p_2 = (1\ 2\ 3)(4\ 6\ 5), \qquad \text{from column 1 to column 2,}$$
$$p_3 = (1\ 3\ 2)(4\ 5\ 6), \qquad\qquad ,, \qquad\qquad ,, \qquad 3,$$
$$p_4 = (1\ 4)(2\ 5)(3\ 6), \qquad\qquad ,, \qquad\qquad ,, \qquad 4,$$
$$p_5 = (1\ 5)(2\ 6)(3\ 4), \qquad\qquad ,, \qquad\qquad ,, \qquad 5,$$
$$p_6 = (1\ 6)(2\ 4)(3\ 5), \qquad\qquad ,, \qquad\qquad ,, \qquad 6.$$

If these permutations form a group then it must be S_3 because p_2 and p_3 are elements of order 3 and p_4, p_5, p_6 are of order 2. It is left as an exercise for the reader to verify that they form a group. A quick way would be to show that the elements harmonise with the generators and relations for D_3, $a^3 = b^2 = e$, $ba = a^2b$.

9.6   Regular permutations

An important feature of permutation groups was noticeable in the last section. Permutations derived from a group table are always *regular*, that is to say that the cycles in any one permutation are all of the same length. For example in p_4 all the cycles are of length 2. The reason for this will emerge in the next section but the reader may feel instinctively that it is connected with the fact that cosets of a subgroup all have the same number of elements.

9.7   Cayley's theorem

This important theorem relates every abstract group to a group of regular permutations and may be stated as follows.

If G is any group of order n then every element in it can be mapped on a regular permutation of degree n and the complete set of these permutations themselves from a group which is isomorphic with G.
Proof. Let A be an arbitrary arrangement of the elements of G. $A = a\ b\ c \ldots x \ldots y \ldots$ up to n elements. If each element in this arrangement is postmultiplied by a particular element x we obtain a

new arrangement which it is reasonable to call Ax.

$$Ax = ax\,bx\,cx \ldots.$$

Similarly $Ay = ay\,by\,cy \ldots,$
and $(Ax)y = (ax)y\,(bx)y\,(cx)y \ldots$
$\qquad\quad = a(xy)\,b(xy)\,c(xy) \ldots$
$\qquad\quad = A\,(xy).$

We can therefore say that for all x, y in G
$$(Ax)y = A(xy). \tag{9.5}$$

Now set out a skeleton table for G, based on A.

Table 9.2

A $\downarrow$	Aa $\downarrow$ a	Ab $\downarrow$ b	Ac $\downarrow$ c	....	Ax $\downarrow$ x	....	Ay $\downarrow$ y	....
a					ax		ay	
b					bx		by	
c					cx		cy	
$\vdots$								
x								
$\vdots$								
y								

The columns of the table are evidently the sequences Aa Ab Ac . . .
Ax . . . Ay . . . as indicated. We can now assign to every element such

as x of G a permutation $\begin{pmatrix} A \\ Ax \end{pmatrix}$ as set out in detail below:

$$p_n = \begin{pmatrix} A \\ Ax \end{pmatrix} = \begin{pmatrix} a & b & c \\ ax & bx & cx \end{pmatrix},$$

$$p_y = \begin{pmatrix} A \\ Ay \end{pmatrix} = \begin{pmatrix} a & b & c \\ ay & by & cy \end{pmatrix}.$$

183

We could describe p_y by saying that its top row consists of all the elements of G in arrangement A while the bottom row is derived from it by postmultiplying every element by y; but we can shift the vertical pairs as we like without altering p_y (see section 9.3). In particular we can change the order of the top row to make it identical with Ax, making corresponding changes in the bottom row, so that the permu-

tation $\begin{pmatrix} A \\ Ay \end{pmatrix}$ reads $\begin{pmatrix} Ax \\ (Ax)y \end{pmatrix}$ and this is the same as $\begin{pmatrix} Ax \\ A(xy) \end{pmatrix}$ from equation

(9.5) above,

i.e.
$$\begin{pmatrix} A \\ Ay \end{pmatrix} = \begin{pmatrix} Ax \\ Axy \end{pmatrix}. \tag{9.6}$$

We are now ready to establish the isomorphism between the set of permutations combined by succession and the group G. The necessary condition is $p_x p_y = p_{xy}$ (see section 2.10 equation (2.1)).

$$\begin{aligned} p_x p_y &= \begin{pmatrix} A \\ Ax \end{pmatrix} \begin{pmatrix} A \\ Ay \end{pmatrix} \\ &= \begin{pmatrix} A \\ Ax \end{pmatrix} \begin{pmatrix} Ax \\ Axy \end{pmatrix} \\ &= \begin{pmatrix} A \\ Axy \end{pmatrix} \\ &= p_{xy}. \end{aligned}$$

If x is an element of order k it is easy to show that p_x breaks up into cycles of order k. Starting with a the first cycle would be $(a\ ax\ ax^2\ \dots\ ax^{k-1})$ then a new element, say d, would produce a cycle $(d\ dx\ \dots)$ and so on until the elements of the group were exhausted. It follows that every p_x is a regular permutation.

9.8 The alternating group A_n

It can be proved, though we shall omit the proof, that in S_n half the permutations are even and that these form a normal subgroup in S_n. The name given to it is A_n the alternating group of degree n. Table 6.4 belongs to a group which is isomorphic with A_4.

9.9 The group of automorphisms of a group

An automorphism of a group can be concisely defined as a permutation of its elements which is also an isomorphism. It was mentioned in section 3.1 that the automorphisms of a group G themselves form a group but the proof was deferred to this chapter. The only condition

that presents any difficulty is that of closure. We established in section
9.4 that permutations are closed under succession so if p_1, p_2 are two
automorphisms p_1p_2 is certainly a permutation of the elements of G.
We must prove in addition that p_1p_2 is an isomorphism.

Let p_1x stand for the image of x under p_1. By the general relation
which characterises all isomorphisms, for any x, y in G

$$p_1(xy) = (p_1x)p_1y \tag{9.7}$$

and $\qquad p_2(xy) = (p_2x)(p_2y) \tag{9.8}$

and replacing x and y in (9.8) by p_1x and p_1y

$$p_2((p_1x)(p_1y)) = (p_2(p_1x))(p_2(p_1y)) \tag{9.9}$$

We shall use (9.7) and (9.9) to prove that

$$p_1p_2(xy) = p_1p_2x\,p_1p_2y,$$

which is the necessary condition for p_1p_2 to be an isomorphism.

$$\begin{aligned}
p_1p_2(xy) &= p_2(p_1(xy)), \text{ by definition of } p_1p_2, \text{ (section 9.4)}\\
&= p_2((p_1x)(p_1y)), \text{ by (9.7)}\\
&= (p_2(p_1x))(p_2(p_1y)), \text{ by (9.9)},\\
&= (p_1p_2x)(p_1p_2y), \text{ by definition of } p_1p_2.
\end{aligned}$$

9.10 Inner and outer automorphisms

As described in section 3.9, the mapping $x \longrightarrow t^{-1}x\,t$ for fixed t produces an automorphism of a non-commutative group which is called
an inner automorphism because it is derived from sources within the
group. Commutative groups have no inner automorphisms except the
trivial identity automorphism but other automorphisms exist for both
commutative and non-commutative groups, and are called *outer
automorphisms*.

If a group G has an outer automorphism p one may conjecture that
there is some larger group K containing G as a subgroup for which p
is part of an inner automorphism. In other words there may be some
element k in K but not in G which plays the part of the transforming
element and that the mapping

$$x \longrightarrow k^{-1}x\,k$$

yields not only an inner automorphism of K but also an outer automorphism of G.

To illustrate these ideas refer again to the group defined by table
3.4 and its sets of transforms in table 3.8. Let K stand for the whole
group and G for the cyclic subgroup $\{1, 2, 3, 4\}$. There are no inner
automorphisms of G but the elements 5, 6, 7, 8 outside G produce not
only some inner automorphisms of K but also an outer automorphism
of G.

$$\begin{array}{ll}
1 \longrightarrow 1 & \quad 3 \longrightarrow 3\\
2 \longrightarrow 4 & \quad 4 \longrightarrow 2
\end{array}$$

9.11 New groups from old

Outer automorphisms provide an interesting way of constructing larger groups from a small one. We shall explain the method by a particular example.

Klein's group is rich in automorphisms for its size. If its elements are e, a, b, c there are six, namely (ab), (ac), (bc), (abc), (acb) and the identity automorphism. Amongst them (ab), (ac), and (bc) are of order 2 while (abc) and (acb) are of order 3. A second-order automorphism may be used to build an 8-group and a third-order automorphism to build a 12-group, each having Klein's group as a normal subgroup N.

Let us decide to use the automorphism (abc) and construct the 12-group G.

(a) Assume the existence of a transforming element g in G but not in N such that

$$g^{-1} a g = b,$$
$$g^{-1} b g = c,$$
$$g^{-1} c g = a.$$

(b) Premultiplying by g these relations lead to

$$\left.\begin{array}{l} ag=gb \\ bg=gc \\ cg=ga \end{array}\right\} \qquad (9.10)$$

and postmultiplying by g converts the first of these into

$$ag^2=gbg, \text{ i.e. } ag^2=g(gc)=g^2c$$

hence

$$\left.\begin{array}{l} ag^2=g^2c \\ bg^2=g^2a \\ cg^2=g^2b \end{array}\right\} \qquad (9.11)$$

Relations (9.10) and (9.11) together with $g^3=1$ are sufficient to define a new group G whose elements listed in cosets are $N=\{e, a, b, c\}$, $A=\{g, ag, bg, cg\}$, $B=\{g^2, ag^2, bg^2, cg^2\}$.

Exercises

9.3   Prove that N commutes with ag and ag^2.

9.4   Find the products $(cg)b$, $(ag)(bg^2)$, and $(bg^2)^2$.

9.5   Find the quotient group of N.

9.6   Find the congruency classes of G.

9.7   Identify G.

9.8   Why is G necessarily a non-commutative group?

9.9   Is G isomorphic with $D_2 \times C_3$?

9.10   Use a method similar to the one above to find an 8-group. Is there more than one 8-group to be found in this way?

THREE SUBSTITUTION GROUPS.
ILLUSTRATIONS OF THE SYMMETRIC GROUP S₄

10.1 A substitution group derived from the field G.F.2³

In section 8.5 we constructed a field G.F.2^3 consisting of eight polynomials of degree 2 over the field Z_2 using an indeterminate R which satisfied the irreducible cubic equation $R^3=R+1$. The seven non-zero elements form a cyclic group under multiplication. This enables us to set out a table of reciprocals among its elements as follows:

Table 10.1 (Reciprocals in G.F.2^3)

$$\begin{array}{ll} 1 & 1 \\ R & R^2+1 \\ R^2 & R^2+R+1 \\ R+1 & R^2+R \end{array}$$

We may also regard R as a different sort of object on which to try the effect of the six transformations discussed in section 2.13.

$$\begin{array}{ll} t \longrightarrow t & t \longrightarrow 1-t \\ t \longrightarrow 1-1/t & t \longrightarrow 1/t \\ t \longrightarrow 1/(1-t) & t \longrightarrow t/(t-1) \end{array}$$

In what follows we shall write the transformation $t \longrightarrow t-1/t$ more shortly as $t-1/t$ and in applying it to R we shall substitute R for t wherever it occurs; similarly for all the other transformations. Taking $*$ as a symbol for substitution

$$\begin{aligned}(1-1/t)*R &= 1-1/R \\ &= 1-(R^2+1), \text{ from table 10.1,} \\ &= R^2.\end{aligned}$$

The result would have been just the same if $1-1/t$ had been changed to $1+1/t$ because in Z_2, 1 and -1 are interchangeable. We may therefore disregard minus signs in the transformations and write them in a more convenient form

Table 10.2

(1) t	(4) t+1
(2) $1+1/t$	(5) $1/t$
(3) $1/(t+1)$	(6) $1+1/(t+1)$

Transformation (6) might equally well be written $t/(t+1)$ but the alternative form is more symmetrical and easier to use.

This set of six applied in turn to R produces the six elements of the field other than 0 and 1. The results in order are

(1) R	(4) $R+1$
(2) R^2	(5) R^2+1
(3) R^2+R	(6) R^2+R+1.

The reader may like to verify that the effect of the six transformations in table 10.2, either singly or in succession is to permute the six R-elements among themselves and that they still form the group D_3 as they did before modification.

In describing the t-expressions of chapter 2 it may be said that the six elements form a group, the rule of combination being the substitution of one expression into another. In the case that we are now considering we must note carefully that this definition breaks down when it is applied to the R-elements. For example if either R or $R+1$ is substituted into R^2+R the result in each case is R^2+R. On the other hand if instead of R^2+R we use the corresponding transformation $1/(t+1)$ we obtain two distinct results

$$1/(t+1)*R=1/(R+1)=R^2+R,$$
$$1/(t+1)*(R+1)=1/R=R^2+1.$$

A careful distinction must therefore be made between the R-elements and the t-transformations. R possesses the property of satisfying the equation $R^3=R+1$ but the group property resides with the transformations.

10.2   A substitution group from G.F.2^5

The group D_3 of transformations was obtained from G.F.2^3 after discarding the elements 0 and 1. Is this a rare accident or are there similar groups related to other Galois fields? Let us consider G.F.2^5. It contains 32 elements, which reduce to 30 on discarding 0 and 1. It can be shown that there are 30 transformations which permute these elements among themselves and which form a group combined as usual by succession. The method used for extracting them is laborious and so the details have been relegated to Appendix 2, but the form of the transformations is interesting. The six listed in table 10.2 appear again as a normal subgroup N with quotient group C_5. The cosets are of the same form as N with t^2, t^4, t^8, t^{16} replacing t. For example one of the cosets is

$$t^2, \; 1+1/t^2, \; 1/(1+t^2), \; (1+t^2), \; 1/t^2, \; 1+1/(1+t^2).$$

The group table is also to be found in Appendix 2 and can be used to identify the group as a direct product group $D_3 \times C_5$. The meaning of the symbols in the table is given in table $A_2.3$ of the appendix. The group will be discussed further in chapter 11 where a suggestion will be put forward for a geometrical representation of it.

10.3   A substitution group from G.F.3³

One more example of this kind of group is given here, obtained from G.F.3³. This contains 27 elements of the form $a R^2 + b R + c$ where R satisfies a cubic equation irreducible over the finite field Z_3, i.e. a, b, c belongs to Z_3 and each may be 0, 1, or 2. A convenient equation for the purpose is

$$R^3 = R + 2.$$

We can take it that the 27 elements form an additive group and that the non-zero elements form a cyclic group for multiplication. The question to consider now is whether, if we discard 0, 1, and 2, there exists a group of twenty-four transformations which permute the remaining twenty-four R-elements among themselves.

The t-transformations of Chapter 2 once more suggest a suitable subgroup: using modulus 3 to abolish minus signs these transformations become respectively

$$\begin{array}{ll} t, & 2t+1, \\ 1+2/t, & 1/t, \\ 1/(2t+1) & 1+1/(t+2). \end{array}$$

The full group emerges as a representation of the symmetric group S_4. Its multiplication table is shown in Appendix 2 at table $A_2.5$ and the meaning of the symbols may be found from table $A_2.4$ which sets out the group of transformations and the set of corresponding R-elements. {A} and {A, B, C} are normal subgroups of the group. The first of these is a Klein's group and the second is the alternating group A_4 described in section 9.8 and used as an exercise in group analysis in section 3.6.

Table $A_2.5$ could also be interpreted as that of all the permutations on four symbols a, b, c, d. One way of matching the permutations with

189

the elements of the table is set out below.

Table 10.3

A	1=I		P	13=(ad)
	6=(ab)(cd)			7=(acdb)
	19=(ac)(bd)			18=(abdc)
	12=(ad)(bc)			24=(bc)

B	2=(abd)		Q	15=(adcb)
	9=(bcd)			5=(ac)
	4=(adc)			8=(abcd)
	11=(acb)			22=(bd)

C	3=(bdc)		R	14=(ab)
	20=(abc)			16=(cd)
	17=(acd)			21=(adbc)
	10=(adb)			23=(acbd)

The isomorphism of the two systems may be verified in particular instances by combining permutations and comparing the result with what might be expected from table $A_2.5$ (note that the product xy in the table means permutation y followed by permutation x). A more general and convincing verification, however, is to apply the method of transforms described in sections 3.9–10. The classes of conjugate permutations revealed by this method speak for themselves.

Exercises

Refer where necessary to table 10.3 and table $A_2.5$ of Appendix 2.

10.1   In table $A_2.5$, $15 \times 3 = 21$. Verify that permutation 15 succeeding permutation 3 is equivalent to permutation 21.

10.2   Find eight elements which form a subgroup of type D_4. How many such subgroups are there? Choose one of them H and find, if possible, another subgroup K such that the extended product of the elements of H and the elements of K form the twenty-four elements of S_4. Could the group be called $D_4 \times C_3$?

10.3   Find all elements conjugate to 13 and all elements conjugate to 7.

10.4   Is D_3 a normal subgroup of S_4?

10.4 Rotations of a cube

Another example of S_4 is the set of rotations of a cube such that it occupies the same cubical space after each rotation from the initial position. It is easy to see that there are 24 different positions of the cube. Each vertex can occupy any of 8 positions and the cube can be rotated through angles of 120° and 240° about the diagonal through that vertex, making 3×8 positions in all. If the symbols a, b, c, d are allotted to the 4 diagonals of the cube then the rotations can be identified with the permutations specified in Table 10.3. Apart from the identity permutation those in set A correspond to half-turns about axes joining the middle points of opposite faces. B and C cover the rotations of 120° and 240° about the four diagonals. The permutations of order 4 in P, Q, R, correspond to quarter and three-quarter turns about the same axes as for set A. Those of order 2 in P, Q, R, correspond to half-turns about lines joining the mid-points of opposite sides of the cube. Typical axes are shown in Fig. 10.1.

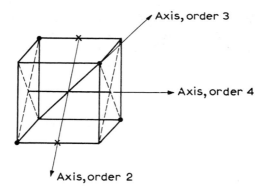

Fig. 10.1

The midpoints of the six faces of the cube may be joined to form an octahedron and therefore the set of rotations for an octahedron like those of the cube form the group S_4. For this reason the group S_4 is sometimes referred to as the octahedral group.

The set of permutations in A, B, C are all even and together they form a subgroup of the group S_4 called the alternating group A_4 of degree 4. In their own right they are isomorphic with the group of rotations of a regular tetrahedron as may be seen from Fig. 10.1 where four corners of the cube have been emphasised to show the tetrahedral pattern.

10.5 Change ringing on bells—plain bob minimus

Bell-ringers discovered how to make use of the permutations of S_4 before groups were ever thought of as a branch of mathematics. On four bells the art consists of 'ringing the changes'—24 in all—without repetition, in conformity with the practical conditions under which bells may be rung. A bell is controlled by wheel and rope and has three basic positions: 'down' at rest, 'up' at handstroke, and 'up' at backstroke.

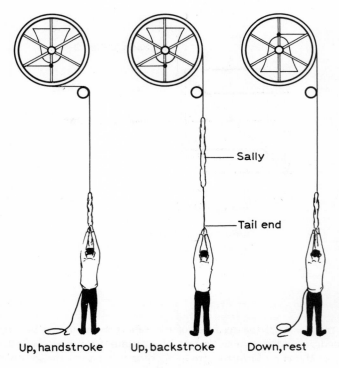

Fig. 10.2

The bell alternates between the two up-positions, the clapper striking the bell just before it reaches the top, so that a perceptible time-interval must elapse between successive blows. While ringing 'rounds', in an unvarying order 1, 2, 3, 4, each bell strikes at regular intervals. To produce 'changes' bells are made to exchange their position in the ringing order in pairs, one ringer holding his bell for a little longer in the balance position at the top, while the other restricts the swing of his bell slightly in order to strike it a little earlier. With four bells 1, 2, 3, 4, a usual sequence called a plain hunt is

```
1   2   3   4
2   1   4   3
2   4   1   3
4   2   3   1
4   3   2   1
3   4   1   2
3   1   4   2
1   3   2   4
─────────────
1   2   3   4
```

So far a regular pattern has been followed: two permutations p and q alternate to produce the eight different arrangements or changes.

p = Interchange the bells in 1st and 2nd places and also those in 3rd and 4th places.

q = Interchange the bells in 2nd and 3rd places.

Note that these are permutations on *places* because in bell-ringing position is all-important. A bell can only exchange with its neighbour. Permutation q involves a change of ringing order between bells in 2nd and 3rd places, not between bell 2 and bell 3. However, symbols for permutations allow either interpretation so long as it is used consistently. Table $A_2.5$ is valid for either places or names if the same interpretation is used for every entry in the table. We may therefore express p as (1 2)(3 4) and q as (2 3). These two used alternately produce the first 8 changes out of 24; then a new permutation r must be used and there are two to choose from (1 2) and (3 4). Since it is customary not to interrupt the course of the treble bell 1, which is now in the lead, (3 4) is chosen as r and produces 1 3 4 2 for the 'lead head' of the next set of 8 changes. This method of ringing the changes on 4 bells is called Plain Bob Minimus and the full course consisting of three 'leads' is

```
1 2 3 4        2 4 3 1

2 1 4 3        4 2 1 3

2 4 1 3        4 1 2 3

4 2 3 1        1 4 3 2
                _____
4 3 2 1        1 4 2 3

3 4 1 2        4 1 3 2

3 1 4 2        4 3 1 2

1 3 2 4        3 4 2 1
_____
1 3 4 2        3 2 4 1

3 1 2 4        2 3 1 4

3 2 1 4        2 1 3 4

2 3 4 1        1 2 4 3
                _____
                1 2 3 4
```

During each lead of 8 changes beginning from the backstroke lead of the treble, all the bells pursue a plain hunt, illustrated in the case of the treble which plain hunts throughout the course. At first this bell 'hunts up', coming in one stroke later on each round, the slower pace suggesting an uphill movement. Then it 'strikes two blows behind', 'hunts down' by quickening its pace and comes into the lead once more. The other three 'working bells' follow a less regular path. At the first backstroke lead of the treble (row 1 3 4 2) bells 2 and 4 are said to 'dodge', while bell 3 'makes second place', i.e. strikes again in the second position. Bell 2 is said to dodge 3–4 down, because it dodges back from 3rd to 4th place while on its downward course after striking two blows behind; in contrast bell 4 is said to dodge 3–4 up, as its dodge comes while it is hunting up. The work of the three bells in the 24 changes from 1 2 3 4 (ringing rounds) back to rounds again comes in the following order:

194

	No. 2	No. 3	No. 4
1st lead	Dodge 3–4 down	Makes seconds place	Dodge 3–4 up
2nd lead	Dodge 3–4 up	Dodge 3–4 down	Makes seconds place
Rounds	Makes seconds place	Dodge 3–4 up	Dodge 3–4 down

Plain Bob Minimus illustrates what is meant by a 'principle of bellringing' defined by the Central Council of Church Bell Ringers:

'A principle is a collection of rows which form a perfect round block, in which:

(i) Every bell does the same work.

(ii) No bell moves up or down more than one place at a time.

(iii) No bell has more than two consecutive blows in any one place'.

This definition is slightly qualified to allow one or more bells to 'plain hunt' (as the treble in Bob Minimus) and to allow 4 consecutive blows in certain places in Doubles, that is changes on 5 bells.

10.6 Plain bob minor

This is an extension of the method used in Plain Bob Minimus and is one of many methods of ringing the 720 changes which are possible on six bells. A plain hunt consists of 12 changes and a plain course consists of 5 leads, each of which is a plain hunt. Two permutations p and q are used alternately and another permutation r called a plain lead end at each backstroke lead of the treble produces a new lead-head.

$$p = (1\ 2)(3\ 4)(5\ 6),$$
$$q = (2\ 3)(4\ 5),$$
$$r = (3\ 4)(5\ 6).$$

The first lead is set out below; the remaining leads can easily be added.

1	2	3	4	5	6		6	5	4	3	2	1
2	1	4	3	6	5		5	6	3	4	1	2
2	4	1	6	3	5		5	3	6	1	4	2
4	2	6	1	5	3		3	5	1	6	2	4
4	6	2	5	1	3		3	1	5	2	6	4
6	4	5	2	3	1		1	3	2	5	4	6

1	3	5	2	6	4

Successive lead heads are:

1	2	3	4	5	6
1	3	5	2	6	4
1	5	6	3	4	2
1	6	4	5	2	3
1	4	2	6	3	5

The treble bell plain hunts throughout the course, while the five working bells vary their paths. The cycle of variations for bell 2 (the first of which comes in row 1 3 5 2 6 4) is:

1st lead of treble	Dodge 3–4 down
2nd lead of treble	Dodge 5–6 down
3rd lead of treble	Dodge 5–6 up
4th lead of treble	Dodge 3–4 up
Rounds	Make seconds place

Bell 3 follows the same cycle but begins by making seconds place.

10.7   Coursing order

If the plain course were set out in full another cyclic arrangement might be noticed. The working bells come to the lead and go up to the back in a definite order 2 4 6 5 3 during each plain hunt. This is called the 'coursing order'. A ringer is acutely aware of this order because it helps him to strike his bell in the correct place. The bell which has its parallel path ahead is his 'course bell' and the one with the following path is his 'after bell'. The five bells as a whole change their position relative to the treble at each lead as a result of permutation r, so that the coursing orders for the six bells during the five leads are respectively:

$$
\begin{array}{cccccc}
1 & 2 & 4 & 6 & 5 & 3 \\
1 & 3 & 2 & 4 & 6 & 5 \\
1 & 5 & 3 & 2 & 4 & 6 \\
1 & 6 & 5 & 3 & 2 & 4 \\
1 & 4 & 6 & 5 & 3 & 2
\end{array}
$$

In order to ring the extent of 720 changes two further permutations are used, a 'bob' which stands for (2 3)(5 6) and a 'single' which stands for (5 6). The conductor may call for either of these to be made instead of r at any backstroke lead of the treble and the effect of each is to change the coursing order of the five working bells. For example, the first lead of the plain course begins with 1 2 3 4 5 6 and ends with 1 3 2 5 4 6, while the coursing order is 1 2 4 6 5 3. What happens in the next lead depends on whether or not the conductor makes a call for a bob or a single. The various effects may be seen in Table 10.4.

Table 10.4

	Next lead head						Coursing order for next lead					
Plain lead end (r)	1	3	5	2	6	4	1	3	2	4	6	5
Bob	1	2	3	5	6	4	1	2	5	4	6	3
Single	1	3	2	5	6	4	1	3	5	4	6	2

It is evident that the coursing order of the five bells is unaffected by r but is altered by either a bob or a single. There are essentially 12 different coursing orders for the five bells. This is half as many as one might perhaps expect but it is to be noted that the order 2 4 6 5 3 arises from the same set of changes as 2 3 5 6 4, which is the same cycle written in reverse order. This may be verified by interchanging the lead head and the lead end in any plain hunt: the rows are the same but in reverse order and the coursing order is also reversed. It would be mathematically possible to ring the 720 changes in a series of 12 plain courses using either a bob or a slightly different kind of single (4 5) to change the coursing order after each plain course; but this would be very monotonous and uninteresting and it is not included among the recognised methods. Plain Bob uses a smaller unit, the plain hunt, as a basis, and Plain Bob Minor resolves the 720 changes into 60 leads corresponding to the 60 different coursing orders which are possible for 6 bells. A 'touch' is defined as any length of changes which is not just a plain course but which starts from and ends with rounds. If we let b stand for a bob and s for a single, then a touch of 360 changes comes from the following sequence repeated twice, three times in all:

$$r\ r\ r\ b\ b\ r\ r\ r\ b\ r.$$

If the very last r is replaced by s then the same triple sequence produces the remaining 360 changes and the final s brings the bells back to rounds. This is one of several methods of ringing the full extent of changes in Plain Bob Minor.

Bobs are usually preferred to singles but at least two singles are necessary to produce all the changes. The reason for this is explained in the next section.

10.8   Odd and even permutations

The need to distinguish between odd and even permutations is important in change ringing. From Table 10.4 it is obvious that r and b produce even permutations of the original coursing order 1 2 4 6 5 3 (see section 10.7). The single produces an odd permutation. It changes an order of 1 2 4 6 5 3 into 1 3 5 4 6 2 and this has already been shown to be the equivalent of the reverse order 1 2 6 4 5 3 so that the effect of s in this example is the single transposition (4 6). This explains why singles are necessary. It is impossible to run through all the 60 coursing orders without resorting to odd permutations.

The idea of odd and even permeates the whole system. p and s are odd permutations of the rows, while q, r, and b are even. Ringers speak of a row being 'in course' or 'out of course' according as it is obtained from rounds by an even or an odd permutation. The rows in the first lead of Plain Bob Minor alternate in pairs following the sequence

e o o e e o o e e o o e

and this pattern is transformed by a single into

o e e o o e e o o e e o.

In the latter case the whole lead is said to be 'out of course' although half the rows are even. A lead which starts from rounds is held to be 'in course' or even.

On the other hand, in Plain Bob Doubles (changes on five bells) the rows in a plain hunt are all even or all odd. The permutations (1 2)(3 4) and (2 3)(4 5) are used alternately and both of them are even. To produce a new lead head either (3 4) or (2 3) is used and since both of these are odd, the plain hunts are alternately 'in course' and 'out of course'.

In Wilfrid Wilson, *Change Ringing* (1965), the author quotes a passage from an earlier classic by C.A.W. Troyte published in 1859: 'On all numbers of bells exactly half the changes are of one nature and half of another; what the nature is, it is out of my power to explain, but as will be seen by-and-by it is a *fact* which must be understood before it is possible to go into the science of composing and proving peals'. Mr. Wilson's book is an authoritative and fascinating account of the Art and Science of Change Ringing in which one may read about many other methods besides Plain Bob. The names themselves, such as Stedman Doubles, Grandsire Triples, Kent Treble Bob, Surprise Royal, and Double Norwich Court Bob Major incite one to enquire further into the mysteries of the belfry.

THREE EXAMPLES OF GROUPS ASSOCIATED WITH REPETITIVE PATTERNS

11.1 Band-patterns as infinite groups

Consider the cyclic pattern below.

Fig. 11.1

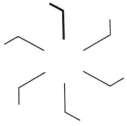

It suggests the finite group C_6 consisting of six rotations clockwise of $0°$, $60°$, $120°$, . . .: any of these applied to the pattern leaves it invariant. The set of transformations might be named 1, T, T^2, T^3, T^4, T^5 or in a slightly different form T^{-2}, T^{-1}, 1, T, T^2, T^3: then T stands for a clockwise rotation of $60°$, T^3 for the same rotation repeated three times and T^{-1} for the inverse rotation of $60°$ anti-clockwise.

The pattern itself might be described as the configuration generated by applying T repeatedly to the motif in Fig. 11.1. After 6 applications it returns to its original position leaving 5 replicas in the intermediate positions.

Now consider an infinite group of transformations

$$\ldots T^{-3},\ T^{-2},\ T^{-1},\ 1,\ T,\ T^2,\ T^3 \ldots$$

applied to the same motif, but this time let T stand for a translation through distance d. The result is an infinite repetitive strip known also as a band-pattern or a frieze. Such patterns are of particular interest to crystallographers and to designers of textiles and wallpapers.

This infinite group is called C_∞ and the pattern is said to be generated from the motif by a single transformation T. Another representation

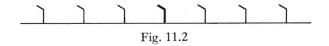

Fig. 11.2

of the same abstract group may be obtained by using a glide-reflection as the sole generator. This is a single transformation compounded from a reflection in a mirror and a translation parallel to the mirror.

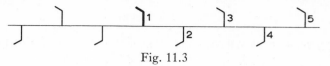

Fig. 11.3

An even number of glide-reflections is evidently equivalent to a translation through a distance equal to twice the glide, while an odd number is equivalent to the same or another glide reflection. The set of equivalent translations takes the motif 1 to positions 3, 5, 7, . . . while the equivalent glide-reflections take it to positions 2, 4, 6, . . . Any of these transformations leave the pattern as a whole invariant.

Using the same motif, a slightly more complicated band may be generated by half-turns h_P and h_Q about one or other of two fixed points P and Q. Following crystallographic notation the symbol $\subset$ (a lens) will be used here to indicate the rotation axis of a half-turn.

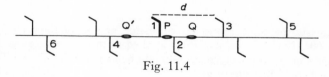

Fig. 11.4

If used alternately starting with h_P the two transformations displace the motif 1 successively to positions 2, 3, 4, 5, 6: position 5 for instance may be reached in four successive steps h_P then h_Q then h_P then h_Q, or in a single step by means of the equivalent translation $1 \longrightarrow 5$. The final transformation in terms of h_P and h_Q is written $h_Q \, h_P \, h_Q \, h_P$ (beginning with the right-hand h_P). The transformation from 1 to 4 is $h_P \, h_Q \, h_P$ and this is equivalent to a half turn $h_{Q'}$ where Q' is the point indicated between 1 and 4. Beginning with h_Q instead of h_P leads successively to all the unnumbered positions in the band. The same pattern could obviously be developed from the motif 1 by using a translation d as generator instead of the second half-turn; but whichever pair of generators is used to make the pattern it is true that the whole band is associated with an infinite set of half-turns about whole sequences of points such as P and Q and also with a set of translations which are integral multiplies of d. The band is invariant under any of these transformations and they are the only transformations which have this property. The group is evidently non-commutative since $h_P \, h_Q \neq h_P \, h_Q$ and it is called D_∞.

200

It is well-known that there are exactly seven essentially different band-patterns which may be generated in this way by a set of geometrical transformations repeated indefinitely in any order. They are listed in Coxeter, *Introduction to Geometry*, p.48, each with suitable generators and the name of its abstract group.

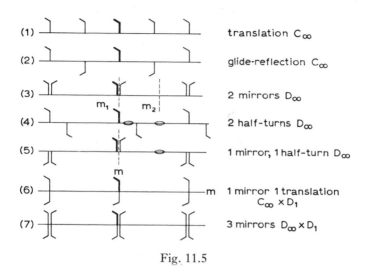

Fig. 11.5

The reason for the name D_∞ in (3) above may not be quite obvious. A useful experiment is to place some object between two mirrors, nearer to one than the other. A pattern of pairs of images will be seen in a circle round the axis of intersection of the mirrors. This is the principle on which a kaleidoscope is constructed. An angle of 30° or $2\pi/12$ between the mirrors produces 11 images which with the original makes 6 pairs, the group associated with the configuration being D_6(c.f.D_4 illustrated in Fig. 3.6). By reducing the size of the angle and allowing the axis to recede to a great distance a circular band approximating to the strip pattern (3) would be obtained. Hence the group associated with this pattern may be regarded as the limit D_∞ to which D_n tends as the centre of the circle recedes and n tends to infinity.

A similar configuration with half-turns about two fixed radii of the circle would lead to band pattern (4). Dihedral groups are not commutative and we have already noted in the description of Fig. 11.4 that $h_P h_Q$ is not the same as $h_Q h_P$. Reflection in the horizontal mirror, a

201

transformation which appears in (6) and (7), commutes with all the other transformations such as half-turns, vertical reflections and translations so these two infinite groups are described as direct products $C_\infty \times D_1$ and $D_\infty \times D_1$.

11.2 Band-patterns as subgroups of a finite group

We shall now look at the band-pattern transformations somewhat differently with a view to combining them into a single finite group. First consider the finite pattern below with 8 elements.

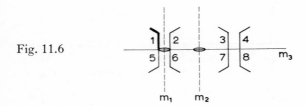

Fig. 11.6

The element 1 may be brought into any one of the 8 positions using transformations as indicated:

(1) $1 \longrightarrow 1$ no change,
(2) $1 \longrightarrow 2$ reflection in a vertical mirror m_1,
(3) $1 \longrightarrow 3$ a translation,
(4) $1 \longrightarrow 4$ reflection in a second vertical mirror m_2,
(5) $1 \longrightarrow 5$ reflection in a horizontal mirror m_3,
(6) $1 \longrightarrow 6$ half-turn,
(7) $1 \longrightarrow 7$ a glide-reflection,
(8) $1 \longrightarrow 8$ a half-turn about a second point.

So far we have not got a group: closure is present for transformations (1), (8), (4), and (5) but the other transformations take some of the elements 1–8 outside the original pattern. However the difficulty disappears if we take as motif not a single element but a whole sequence of elements all of which are to receive the label 1: this is analogous with treating the integers 1, 4, 7, 10, . . . as members of the congruence class 1 (mod 3).

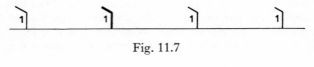

Fig. 11.7

202

Now make a complete band-pattern by adding the congruence classes 2–7.

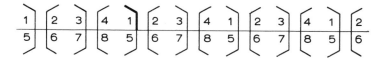

Fig. 11.8

Consider the effect of the 8 transformations on this pattern and particularly the effect of (2), (3), (6), and (7) which before did not conform to the axiom of closure. Transformation (3) sends every 1 to a 3 and every 3 to a 1. It also produces interchanges $2\longleftrightarrow4$, $5\longleftrightarrow7$, $6\longleftrightarrow8$. Other results are:

$$(2) \quad 1\longleftrightarrow2, 3\longleftrightarrow4, 5\longleftrightarrow6, 7\longleftrightarrow8;$$
$$(6) \quad 1\longleftrightarrow6, 2\longleftrightarrow5, 3\longleftrightarrow8, 4\longleftrightarrow7;$$
$$(7) \quad 1\longleftrightarrow7, 2\longleftrightarrow8, 3\longleftrightarrow5, 4\longleftrightarrow6.$$

The structure of the finite group $C_2\times C_2\times C_2$ emerges very clearly (see section 3.11): all the transformations except the identity are of order 2. There are 7 subgroups of order 2 and also 7 of order 4 (all Klein's), their respective elements being:

(1)	(2)	(3)	(4)
(1)	(2)	(5)	(6)
(1)	(2)	(7)	(8)
(1)	(3)	(5)	(7)
(1)	(3)	(6)	(8)
(1)	(4)	(5)	(8)
(1)	(4)	(6)	(7)

Each of these fourteen subgroups, easily recognisable in Fig. 11.6, corresponds to one or other of the 7 band-patterns already listed in Table 11.1; no new ones emerge. For example, the subgroup {(1), (3)} yields . . . 1 3 1 3 . . . which is essentially the same as that produced by the identity group . . . 1 1 1 . . . Again the subgroups (1) (6), (1) (8) and (1) (3) (6) (8) all yield the same kind of band, number (4) in Fig. 11.5.

11.3　A card problem

Is it possible to arrange the sixteen court cards of a pack in a four-by-four array which complies with both of the conditions below?

1.　Each row and column is to contain an ace, a king, a queen and a jack.
2.　Each row and column is to contain one of each suit.

This is a challenging problem which often arouses keen interest in anyone asked to solve it. Sooner or later a solution is found, usually by trial and error. Any reader who has not met this problem before may wish to try his skill before reading any further. In any case a pack of cards, and preferably two packs, is almost indispensable for pursuing the further lines of investigation, which open up once a successful arrangement has been found and noted; for example these three:

(a) How many different arrangements of the cards meet the two conditions?

(b) What permutations of the cards or the places will transform one arrangement into another?

(c) Do all the permutations of this kind form a group? If so can the group be identified?

11.4　Permutations

It is perhaps best to begin with question (b) as a stepping-stone to answering question (a). The reader will soon convince himself that any of the following classes of permutations will successfully permute the cards.

(1)　Rearrangement of whole rows—class R.

(2)　Rearrangement of whole columns—class C.

(3)　Permutations of suits, e.g. interchanging the aces of hearts and diamonds, A_H and A_D and also K_H and K_D, Q_H and Q_D, J_H and J_D— class S.

(4)　Permutations of denominations of cards, e.g. interchanging kings and queens in their respective suits or making a cyclic change, kings to queens, queens to jacks, jacks to queens—class D.

11.5　What do we mean by different arrangements?

There are two ways of looking at this question. One is the obvious way in which if two arrangements of the cards are compared they are considered different if any card is seen to occupy two different positions.

Another and more profitable way is to regard the sixteen cards as an element in a repetitive pattern occupying an infinite plane space. Then the two arrangements below though apparently different would in fact be indistinguishable.

A_S	J_C	K_H	Q_D
J_D	A_H	Q_C	K_S
K_C	Q_S	A_D	J_H
Q_H	K_D	J_S	A_C

Fig. 11.9

Q_C	K_S	J_D	A_H
A_D	J_H	K_C	Q_S
J_S	A_C	Q_H	K_D
K_H	Q_D	A_S	J_C

Fig. 11.10

The two arrangements overlap in the same infinite pattern.

Fig. 11.11

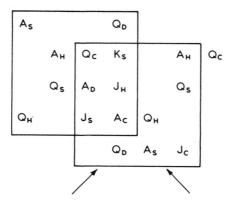

Note that the diagonals of aces and queens clearly seen in Fig. 11.9 are still implied in Fig. 11.10 and can easily be recognised along the lines indicated by arrows.

To give the array an infinite setting reduces the possible number of different arrangements by the factor 16 since any selected card may occupy any one of sixteen positions without the basic pattern being altered. Any four-by-four array of the system can be regarded as one of an equivalence class of sixteen members. We shall always take the array in which the ace of spades occupies the N.W. corner (north at the top of the page) as being the standard representative of its class.

We can now proceed to investigate the number of distinct arrangements which have the ace of spades in this position. Fig. 11.9 showed one such arrangement. It has been carefully chosen as the initial arrangement a_i for the following reasons: the aces run down one of the diagonals in the sequence spades, hearts, diamonds, clubs, memorable to many people as the ranking order of suits in contract bridge; the spade suit runs down the rows in the order A K Q J and could be made to form a diagonal by commuting the columns.

However a_i is not the only arrangement to possess these features (the reader may like to look for other possible arrangements and find their number).

11.6 Permutations of columns

Starting from a_i it is interesting to see what other patterns appear if the columns are permuted in the six ways which are possible, allowing for the ace of spades being fixed in the first column. There are three distinctive diagonal patterns in the infinite array which appear as the columns are permuted.

(i) Aces and kings run N.W.–S.E in the chosen cyclic order of suits but in opposite directions. Queens and jacks are parallel and opposite in the N.E.–S.W. direction.

(ii) Suits are parallel in pairs following in some direction the chosen order A K Q J.

(iii) Parallel pairs of lines contain sequences in the order A K Q J, the four cards being in different suits.

Each of the three patterns has two versions, the directions N.W.–S.E. and N.E.–S.W. being interchangeable.

11.7 Number of distinct arrangements

We can now count all the possible arrangements having the ace of spades in the N.W. corner by examining the choices open to us in arriving at the initial arrangement a_i.

(a) Choose one of the six patterns mentioned above and decide to have aces in the diagonal N.W.–S.E. *This is one of six choices.*

(b) Next arrange the aces in the order S H D C. With spades fixed this is also *one of six choices*.

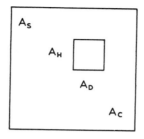

Fig. 11.12

(c) Now decide that the spades are to appear down the rows in the order A K Q J, again, *one of six choices*.

(d) To comply with the rules the empty box in Fig. 11.12 must be filled by a black king, queen, or jack. From (c) the king of spades must go in the box or to the right of it. If we put him in the box the next step is to complete the diagonal of kings following the same order as the aces. The rest of the pattern is then determined. If we place him to the right of the box, then we must put either Q_C or J_C in the box. Only Q_C leads to the correct sequence A K Q J of spades, and the pattern when completed becomes a_i. Hence Q_C is *one of two choices* for the empty box.

Altogether then we appear to have $6 \times 6 \times 6 \times 2$ choices in laying out the cards in the form a_i, and this makes a total of 432 arrangements. If so there must be 432 permutations of a_i which will transform it into some permissible array.

11.8 Do the permutations form a group? If so which group?

We have already noted four small groups of type S_3 (the symmetric group of permutations on three symbols, isomorphic with D_3). They are the groups R, C, S, and D defined as classes in section 11.3. This suggests the presence of a main group G of order 432 having the others as subgroups. Experiment also suggests that the small groups act independently like the component groups of a direct product group. It is therefore advisable at this point to establish whether or not each element of any of R, C, S, D commutes with *all* elements of the other three groups.

207

With any four-by-four arrangement it can be seen at once that if permutations c and r are elements respectively of C and R then

$$cr=rc.$$

For much the same reasons it is also true that

$$sd=ds,$$

a statement which would obviously be true if the cards were arranged

A K Q J	spades
A K Q J	hearts
A K Q J	diamonds
A K Q J	clubs

while the added complication of the cards being distributed less regularly in no way affects the commutative property of d and s.

It is rather less obvious that c and s commute especially as c is a permutation of places while s is a permutation of cards (see section 10.5). Suppose that s is an exchange between hearts and diamonds (HD) while c moves the second column to fourth, fourth to third, and third to second (243). Then if we examine the fates of A_H and A_D, positioned at random as in Fig. 11.13 below, both sc and cs lead to the same final result $A_H{}' A_D{}'$.

Fig. 11.13

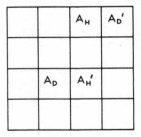

An analogy may be helpful here. If two officers O_1 and O_2 are posted to depots D_1 and D_2 where units are about to embark for foreign service at stations S_1 and S_2, and if through a clerical error O_1 goes to D_2 and O_2 to D_1 there are two ways, equally efficacious, of retrieving the mistake. The officers can be flown between depots and then transported to the right stations or they can be transported to the wrong stations and change stations on arrival. Both lead to the right result so the two operations of exchanging officers and transportation from depots to stations are commutative.

We now have good reason to suspect that the elements of the three groups S, D, and C (excluding R *pro. tem.*) together with all their products may form a group H isomorphic with $S \times D \times C$. If so its order would be 216, half that inferred in section 11.5. Also every element of the group would be expressible in the form sdc. To test this in a random case take the initial arrangement a_i and obtain two new arrangements from it by using the two permutations $s_1 d_1 c_1$ and $s_2 d_2 c_2$ where

$$s_1 = \text{(clubs diamonds)}, \qquad s_2 = \text{(hearts clubs)},$$
$$d_1 = \text{(KQ)}, \qquad d_2 = \text{(KQJ)},$$
$$c_1 = \text{(243)}, \qquad c_2 = \text{(23)}.$$

The resulting arrangements a_x, a_y are given below.

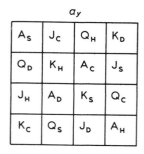

	a_x		
A_S	Q_H	K_C	J_D
J_C	K_D	Q_S	A_H
Q_D	A_C	J_H	K_S
K_H	J_S	A_D	Q_C

Fig. 11.14

	a_y		
A_S	J_C	Q_H	K_D
Q_D	K_H	A_C	J_S
J_H	A_D	K_S	Q_C
K_C	Q_S	J_D	A_H

Fig. 11.15

The question now is whether it is possible to get from a_y to a_x by a permutation of the form sdc. This is readily seen to be possible.

(HCD) puts the aces in the right order SHCD;

(QJ) puts the spades in the right order AQKJ;

(34) is then a column shift which achieves a_x as a final arrangement.

Other values for a_x and a_y will also be found capable of being transformed into one another by a permutation of the form sdc.

Another way of testing is to transform a_i into another arrangement by a transformation of the *rows*, that is to say an element of the group R which up to now we have excluded, and then to see whether this can be transformed back into a_i by an element sdc. For example transposition of the middle two rows of a_i leads to the arrangement of Fig. 11.16.

209

Let us call this a_z. Proceeding as before:

(HD) rearranges the aces,

(KQ) rearranges the spades,

(23) puts the aces into the diagonal,

but this time we arrive not at a_i but the arrangement in Fig. 11.17.

A_S	J_C	K_H	Q_D
K_C	Q_S	A_D	J_H
J_D	A_H	Q_C	K_S
Q_H	K_D	J_S	A_C

Fig. 11.16

A_S	Q_D	J_C	K_H
Q_C	A_H	K_S	J_D
J_H	K_C	A_D	Q_S
K_D	J_S	Q_H	A_C

Fig. 11.17

This arrangement may be recognised as the one we should have obtained if we had put the king of spades in the empty box (Fig. 11.12). It is the last rejected choice in section 11.5.

Further experiments with the elements of R show that any odd permutation of the rows, namely (23), (34) or (24) produces an arrangement like a_z which cannot be transformed into a_i by any element of type sdc. On the other hand the even permutations consisting of (234), (243), and the identity permutation produce arrangements like a_x and a_y.

11.9 The complete group

If all 216 arrangements such as a_x developed from a_i by the group H are transformed by an odd permutation r_o of the group R then we get the remaining 216 arrangements of the total number already determined as 432. Let us call these the set of odd arrangements A_o. The coset of the subgroup H is r_oH. An even permutation r_e of H produces A_e the set of even arrangements. If we were confronted with any two arrangements and were asked to specify the permutation which would convert one into the other we should want to know whether or not the two arrangements belonged to the same set A_e or A_o. It would be better still if we could decide at a glance whether any arrangement is odd or even. Fortunately a simple criterion can be found. In a_i which belongs to A_e there is a diagonal of aces crossing one of queens and these follow the same order of suits down the rows S H D C. In any odd arrangement the order of suits for the two diagonals is opposite as in Fig. 11.17. We shall shortly delve into the reason for this but if we now assume it to be true the fact will enable us to determine whether an arrangement is odd or even. In Fig. 11.17 it is easy to see that the diagonal of kings runs up the page in the order S H D C while the aces run down in that order. This indicates that it is an odd arrangement. In Fig. 11.14 however it must first be determined whether kings, queens, or jacks form the opposite diagonal. The king of clubs tells us that it is again a diagonal of kings because this is the card in the top row above the bottom ace A_D. Kings run down in the same order S H C D as the aces so the arrangement is even.

The reason for this criterion being valid emerges from consideration of the effect on an even arrangement such as a_i of the various permutations of rows. They can be summarised briefly.

Odd permutations

(23) This permutation (a) after a column change (23) puts the queens in the diagonal opposite the aces,

      (b) changes the order of aces from S H D C to S D H C,

      (c) changes the order of queens from D C S H to D S C H.

(34)   (a) Puts the kings in opposite diagonal,

      (b) changes aces from S H D C to S H C D,

      (c) changes kings from H S C D to H S D C.

(24)   (a) Puts jacks in opposite diagonal,

      (b) changes aces from S H D C to S C D H,

      (c) changes jacks from C D H S to C S H D.

In each case the opposite diagonals now run in opposite directions.

Even permutations

A similar analysis for even permutations show that each of them leads to a pair of diagonals having the same order of suits down the rows. There is now considerable justification for concluding that the complete group G arises from four smaller groups SDC and R′ where R′ consists of the identity permutation and one of the odd permutations of the rows. G is isomorphic with $S_3 \times S_3 \times S_3 \times C_2$. Since R and C are of the same nature it could evidently be derived equally well from S, D, C′, and R and from the essential symmetry of the system probably also from S′, D, C, R or S, D′, C, R.

To convert from any arrangement a_p to another a_q convenient steps would be

(a) decide whether a_p and a_q are odd or even;
(b) if different from each other transpose a pair of rows.
(c) permute suits to get the aces in the right order;
(d) permute denominations to get the spade suit in the right order;
(e) permute columns as necessary.

11.10   An example of a space-group

The group obtained in section 10.2, which we shall now call G, has some interesting features and its complete table is displayed at the end of this chapter (Table 11.3). It is arranged on the cyclic subgroup H generated by the transformation $1+1/D^2$. Since it consists of half the elements it must commute with the other half, the coset L, and so is a normal subgroup. It was also observed in section 10.2 that the six key elements (1), (11), (21), (2), (12), (22) made a normal subgroup of type D_3. At first sight the whole group might possibly be of type D_{15} but an investigation of the order of the even-numbered elements shows that this is not the right name for it. In a dihedral group all the elements in the coset should be of order 2; whereas in L all but (2), (12), and (22) are of order 10. There are three subgroups of order 10: omitting brackets, while remembering that the numbers stand for D-transformations and not R-elements, they are

A	1	6	19	24	7	12	25	30	13	18,
B	1	16	19	4	7	22	25	10	13	28,
C	1	26	19	14	7	2	25	20	13	8.

The elements 1, 19, 7, 25, 13 stand out as a subgroup K common to A, B, and C; they are also a subgroup of the larger subgroup H. Readers may like to confirm that K is a normal subgroup and that the group G specified by Table 11-3 is of type $D_3 \times C_5$.

A is not a normal subgroup but we can usefully partition the group G into its left cosets by multiplying A on the left, first by 3 and then by 5, using the table to help us.

Table 11.1

A	1	6	19	24	7	12	25	30	13	18,
A_1	3	28	21	16	9	4	27	22	15	10,
A_2	5	20	23	8	11	26	29	14	17	2.

Different starting points in A_1 and A_2, and the same cyclic order, produce a more symmetrical arrangement which harmonises with the subgroups A, B, C.

Table 11.2

A	1	6	19	24	7	12	25	30	13	18,
A_1	21	16	9	4	27	22	15	10	3	28,
A_2	11	26	29	14	17	2	5	20	23	8.

The triplets (1, 11, 21) (2, 12, 22) etc., are now conspicuous and five of these triplets are also visible as corresponding elements in the subgroups A, B, C.

Consider now the effect of any even transformation on the array of elements in Table 11.2: for example transformation 4 has this effect

$$1 \longrightarrow 4 \qquad 6 \longrightarrow 27$$
$$21 \longrightarrow 24 \qquad 16 \longrightarrow 7$$
$$11 \longrightarrow 14 \qquad 26 \longrightarrow 17$$

and so on. The transformation moves every element three places to the right and also interchanges rows A and A_1:

6 moves 1 place and interchanges A_1 and A_2;

8 moves 9 places (or 1 to the left) and changes A and A_1;

10 moves 7 places and changes A and A_2.

Every even transformation follows the same general pattern, a displacement common to all elements and interchange between two of the three rows.

These considerations suggest a space-lattice which might be amenable to this set of transformations. First consider a two-dimensional triangular lattice of points numbered 1, 11, or 21 such that every triangle has these three numbers for its vertices in some order.

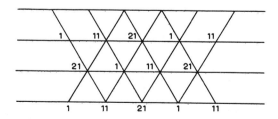

Fig. 11.18

213

A half-turn about any point 1 leaves the set of points numbered 1 unchanged as a whole, though not individually, and converts every 11 to 21 and vice-versa. The same is true for rotations $\pi/3$ or $-\pi/3$. Similar rotations about a point marked 11 interchange the 1's and 21's and have no effect on the lattice of 11's.

We can now envisage a three-dimensional model consisting of ten such lattices placed one above the other following the order suggested by A, A_1, A_2. Each layer consists entirely of one of the triplets (1, 11, 21), (6, 26, 16), ...

Fig. 11.19

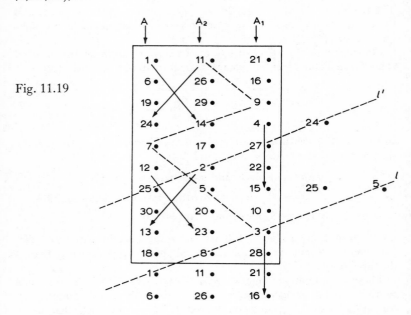

A *Screw* is a three-dimensional transformation which, like a glide reflection or a roto-reflection consists of two transformations compounded. The components are a translation and a rotation about an axis parallel to the translation. All the even transformations in the group G can be regarded as screws having vertical axes through one or other of the typical points 1, 11, 21. If we define 6 as a rotation of $\pi/3$ about axis 1 and a displacement d, i.e. $(1, \pi/3, d)$, then, following the subgroup A, $19 = (1, 2\pi/3, 2d)$, $24 = (1, \pi, 3d)$ etc.

In Fig. 11.19 the arrows show the effect of screw number 14 $(21, \pi, 3d)$ on 6 of the 30 elements. Every element moves down through 3 layers from its original position and elements in A and A_2 change over. This array accounts for the action of all the even transformations but there

214

Table 11.3

	H															L														
	1	3	5	7	9	11	13	15	17	19	21	23	25	27	29	2	4	6	8	10	12	14	16	18	20	22	24	26	28	30
H 1	1	3	5	7	9	11	13	15	17	19	21	23	25	27	29	2	4	6	8	10	12	14	16	18	20	22	24	26	28	30
3	3	9	15	21	27	2	8	14	20	26	1	7	13	19	25	6	12	18	24	30	5	11	17	23	29	4	10	16	22	28
5	5	15	25	4	14	24	3	13	23	2	12	22	1	11	21	10	20	30	9	19	29	8	18	28	7	17	27	6	16	26
7	7	21	4	18	1	15	29	12	26	9	23	6	20	3	17	14	28	11	25	8	22	5	19	2	16	30	13	27	10	24
9	9	27	14	1	19	6	24	11	29	16	3	21	8	26	13	18	5	23	10	28	15	2	20	7	25	12	30	17	4	22
11	11	2	24	15	6	28	19	10	1	23	14	5	27	18	9	22	13	4	26	17	8	30	21	12	3	25	16	7	29	20
13	13	8	3	29	24	19	14	9	4	30	25	20	15	10	5	26	21	16	11	6	1	27	22	17	12	7	2	28	23	18
15	15	14	13	12	11	10	9	8	7	6	5	4	3	2	1	30	29	28	27	26	25	24	23	22	21	20	19	18	17	16
17	17	20	23	26	29	1	4	7	10	13	16	19	22	25	28	3	6	9	12	15	18	21	24	27	30	2	5	8	11	14
19	19	26	2	9	16	23	30	6	13	20	27	3	10	17	24	7	14	21	28	4	11	18	25	1	8	15	22	29	5	12
21	21	1	12	23	3	14	25	5	16	27	7	18	29	9	20	11	22	2	13	24	4	15	26	6	17	28	8	19	30	10
23	23	7	22	6	21	5	20	4	19	3	18	2	17	1	16	15	30	14	29	13	28	12	27	11	26	10	25	9	24	8
25	25	13	1	20	8	27	15	3	22	10	29	17	5	24	12	19	7	26	14	2	21	9	28	16	4	23	11	30	18	6
27	27	19	11	3	26	18	10	2	25	17	9	1	24	16	8	23	15	7	30	22	14	6	29	21	13	5	28	20	12	4
29	29	25	21	17	13	9	5	1	28	24	20	16	12	8	4	27	23	19	15	11	7	3	30	26	22	18	14	10	6	2
L 2	2	6	10	14	18	22	26	30	3	7	11	15	19	23	27	4	8	12	16	20	24	28	1	5	9	13	17	21	25	29
4	4	12	20	28	5	13	21	29	6	14	22	30	7	15	23	8	16	24	1	9	17	25	2	10	18	26	3	11	19	27
6	6	18	30	11	23	4	16	28	9	21	2	14	26	7	19	12	24	5	17	29	10	22	3	15	27	8	20	1	13	25
8	8	24	9	25	10	26	11	27	12	28	13	29	14	30	15	16	1	17	2	18	3	19	4	20	5	21	6	22	7	23
10	10	30	19	8	28	17	6	26	15	4	24	13	2	22	11	20	9	29	18	7	27	16	5	25	14	3	23	12	1	21
12	12	5	29	22	15	8	1	25	18	11	4	28	21	14	7	24	17	10	3	27	20	13	6	30	23	16	9	2	26	19
14	14	11	8	5	2	30	27	24	21	18	15	12	9	6	3	28	25	22	19	16	13	10	7	4	1	29	26	23	20	17
16	16	17	18	19	20	21	22	23	24	25	26	27	28	29	30	1	2	3	4	5	6	7	8	9	10	11	12	13	14	15
18	18	23	28	2	7	12	17	22	27	1	6	11	16	21	26	5	10	15	20	25	30	4	9	14	19	24	29	3	8	13
20	20	29	7	16	25	3	12	21	30	8	17	26	4	13	22	9	18	27	5	14	23	1	10	19	28	6	15	24	2	11
22	22	4	17	30	12	25	7	20	2	15	28	10	23	5	18	13	26	8	21	3	16	29	11	24	6	19	1	14	27	9
24	24	10	27	13	30	16	2	19	5	22	8	25	11	28	14	17	3	20	6	23	9	26	12	29	15	1	18	4	21	7
26	26	16	6	27	17	7	28	18	8	29	19	9	30	20	10	21	11	1	22	12	2	23	13	3	24	14	4	25	15	5
28	28	22	16	10	4	29	23	17	11	5	30	24	18	12	6	25	19	13	7	1	26	20	14	8	2	27	21	15	9	3
30	30	28	26	24	22	20	18	16	14	12	10	8	6	4	2	29	27	25	23	21	19	17	15	13	11	9	7	5	3	1

are still the odd ones to consider. Does the model also accommodate these? We need a transformation 3 to produce the cyclic effect

$$1 \longrightarrow 3 \longrightarrow 5 \longrightarrow 7 \ldots ,$$
$$2 \longrightarrow 24 \longrightarrow 16 \longrightarrow 8 \ldots .$$

An obvious candidate is a translation parallel to the lines 1, 1' in the array above but it would be more in keeping with the other transformations to choose a screw about another vertical axis, this time through the *centre* of one of the triangles instead of through a vertex. This would produce the sequence $1 \longrightarrow 3 \longrightarrow 5$ shown by the dotted zigzag line in the diagram causing a cyclic interchange of columns A, A_1, A_2 with a translation $2d$ upwards. Each of the 30 elements is now identified with a screw, including of course the zero screw which produces no change.

Towards the end of this book we come back to the beautiful mathematical idea of isomorphism. It is exhilarating to start from an abstract algebraic structure derived from an irreducible equation and a finite field of polynomials and arrive finally at a set of screw-transformations in 3-space, under the guidance of a group which connects the two systems apparently so remote from each other.

Although we have paid some attention to rings and fields, our chief concern has been with the structure of finite groups and with the sometimes surprising connection between two apparently different mathematical systems which happen to be isomorphic. The concluding chapter will discuss some general ideas on groups suggested by these studies.

12

WHAT IS A GROUP?

I remember as a child being highly amused by an anecdote about a traveller in Ireland: being completely lost he enquired of a passer-by how to find his way to Ballymoy. He received first a dismayed look and then the reply 'Ah now, if you want to go to Ballymoy you shouldn't start from here!'

I have somewhat the same sense of dismay when I consider the axioms for a group which are often, but not always, presented as the starting-point for any study of this subject. Standard axioms are set out in Appendix 1 and, omitting some of the details, the raw material for a group is said to be a set of elements $a, b, c, \ldots$ together with something symbolised by $*$ and called either a rule of combination or a *binary operation*, meaning a method of combining two of the elements. With this formula the axioms neatly telescope two radically different situations among real representations of abstract groups:

1. The set $a, b, c, \ldots$ consists of inert elements and $*$ represents an active operation, a typical example being the set of residues to a modulus for the operation of addition.

2. The set consists of active operations while $*$ stands for succession, one operation following on another, as in the case of rotations of a square.

The use of the same word operation for completely different components of the group structure is a potent source of confusion and I suspect that mainly for this reason the axioms, being a very sophisticated formulation of the rules governing both situations, may sometimes act as a barrier rather than as a gateway to initial understanding. Now that group theory is beginning to enter the school curriculum, even at quite an early stage, the question of finding an approach suitable for young minds becomes very important. Suppose we take the view that the first essential is for them to study a variety of group representations and leave the consideration of precise axioms till a later stage. Suppose further that the course begins with geometrical transformations and goes on to permutations of various objects such as cards or letters of the alphabet. Some such ideas as the following are likely to be developed as the course proceeds:

(a) Each group contains active and passive elements, an equal number of each. (It would be observed that the no-change or standstill transformation always insists on being included among the active elements).

(b) The passive elements are a democracy: any or all of them are equally suitable for demonstrating the effect of the active elements. These, on the other hand, are highly individual, each playing a distinctive part in the organisation of the group, although some subsets may perhaps be distinguished whose members have similar rôles.

(c) Two transformations combined in succession always have the same effect as some single transformation; and it is exceptional for the combination to be commutative.

Synthesising these ideas by means of group tables it would appear that the property of being a group resides with the set of active elements and that there is a universal rule of combination, namely succession.

If now a different kind of representation, such as that of residue classes, were introduced, would it fit into this picture? Have we still got active and passive elements and does the rule of succession still apply? It is possible to answer yes to both questions. The active and passive elements are still inherent in the system but so close is the affinity between them that their separate existence is not usually recognised. Taking the number 3 as prototype modulus, the passive elements are the residue classes 0, 1, 2, while the active operators are (0), (1), (2), standing respectively for Add 0, Add 1, Add 2. Any two of these operators applied in succession to any of the elements 0, 1, 2 have the same effect as a single operator, the group table being:

S	(0)	(1)	(2)
(0)	(0)	(1)	(2)
(1)	(1)	(2)	(0)
(2)	(2)	(0)	(1)

It is a very short step from this to the table which is usually set out for the passive elements:

+	0	1	2
0	0	1	2
1	1	2	0
2	2	0	1

Unlike the geometrical transformations and permutations the operators in the modular arithmetic are all of one kind and it is this which makes it possible to adopt addition as the rule of combination. Both tables fulfil the conditions required by the axioms and they illustrate the distinction made under (1) and (2) in this chapter. This is perhaps the reason why in such structures as rings and fields which contain addition and multiplication groups as substructures, it is natural to think of the elements of the ring or field as passive. The distributive property $a(b+c)=ab+ac$ makes most sense if we regard a, b, and c as objects of manipulation by the active operations of addition and multiplication; but when groups are considered in their own right I have a strong conviction that both active and passive elements are essential components and that they always exist though sometimes not very evident. The t-expressions of Chapter 2 are another instance of extremely close affinity between the two sets. They are identical in form but not in function. The axioms in this case conceal rather than display the possibilities of the situation. They direct attention to six elements and one operation, substitution, and there is little room for manoeuvre when the situation is expressed in this way. Distinguishing between the active and passive sets makes it possible to keep the transformations constant while varying the meaning to be attached to the expressions on which they operate. Developments of this kind were described in Chapters 2 and 10. Mathematicians and scientists are necessarily concerned with reducing the axioms for an algebraic structure to the barest minimum but there is a danger of something being lost in the process if in a particular representation no account is taken of the dual sets of elements.

Historically groups arose as groups of transformations: agreement on the existing formal axioms did not come until long after groups were discovered and used. This is another reason why a study of transformations might make a more suitable starting point for a school course than the axioms. The road to Ballymoy or any distant goal may be easier for the traveller if he begins from the most favourable place!

APPENDIX 1

DEFINITIONS

A$_1$.1 Structures

1. *Group*

A group G is a set of elements a, b, c, . . . with a rule of combination $*$ such that:

(i) $a*b$ is in G for all a, b in G;

(ii) $(a*b)*c=a*(b*c)$ for all a, b, c in G;

(iii) there is a special element e, called the identity element, in G such that for all a in G

$$e*a=a \text{ and } a*e=a;$$

(iv) for each a in G there is an element a^{-1}, called the inverse of a, in G such that

$$a*a^{-1}=e \text{ and } a^{-1}*a=e.$$

In a commutative group $a*b=b*a$ for all a, b in G.

2. *Ring*

A ring R is a set of elements a, b, c, . . . with two rules of combination $+$ and $\cdot$ such that:

(i) R is a commutative group with respect to $+$, with its identity element written as 0 and the inverse element of a written as $-a$;

(ii) for all a, b in R $a \cdot b$ is in R;

(iii) $(a \cdot b) \cdot c=a \cdot (b \cdot c)$ for all a, b, c in R;

(iv) $a \cdot (b+c)=(a \cdot b)+(a \cdot c)$ and $(a+b) \cdot c=(a \cdot c)+(b \cdot c)$ for all a, b, c in R.

(Throughout the text of this book $a \cdot b$ is written as ab).

3. *Integral domain*

An integral domain D is a ring with three further properties:

(i) It has an identity element 1 for the second rule of combination.

(ii) $a \cdot b = b \cdot a$ for all a, b in D.

(iii) If $a \cdot b = 0$ then either $a = 0$ or $b = 0$.

4. *Field*

A field F is a ring of elements a, b, c, ... in which the non-zero elements form a commutative group with respect to the second rule of combination. (It has one more property than the integral domain, that of possessing an inverse element a^{-1}, in respect to $\cdot$, for each a in F).

A_1.2 Equivalence relation on a set

An equivalence relation R on a set S is such that:

(i) a R a for all a in S (reflexive);

(ii) if a R b then b R a for all a, b in S (symmetric);

(iii) if a R b and b R c then a R c for all a, b, c in S (transitive).

A_1.3 Isomorphism

An isomorphism between two groups $(G *)$ and $(H \otimes)$ is a one-one mapping $a \longleftrightarrow \varphi(a)$ where a is in G and $\varphi(a)$ is in H such that for all a, b in G

$$\varphi(a * b) = \varphi(a) \otimes \varphi(b).$$

Similarly an isomorphism between two rings $R (+\cdot)$ and $R' (\oplus \odot)$ is a one-one mapping $a \longrightarrow \varphi(a)$ where a is in R and $\varphi(a)$ is in R' such that for all a, b, in R

$$\varphi(a + b) = \varphi(a) \oplus \varphi(b)$$

and $$\varphi(a \cdot b) = \varphi a \odot \varphi(b).$$

APPENDIX 2

SUBSTITUTION GROUPS
DERIVED FROM GALOIS FIELDS

This appendix supplies the details of the method used for finding substitution groups from each of two Galois fields G.F.2^5 and G.F.3^3, described in Chapter 10.

A$_2$.1 The group of transformations derived from G.F.2^5

There are 32 elements in the field. The non-zero elements, all polynomials in R of degree four or less, form a cyclic group of order 31 for multiplication. A table of powers of R may be found, using an irreducible quintic equation such as $R^5 = R^2 + 1$.

In table A$_2$.1 below each power is put opposite its reciprocal. (Note that $R^{31} = 1$).

Table A$_2$.1

$$R^1 = R \qquad\qquad R^{30} = R^4 + R$$
$$R^2 = R^2 \qquad\qquad R^{29} = R^3 + 1$$
$$R^3 = R^3 \qquad\qquad R^{28} = R^4 + R^2 + R$$
$$R^4 = R^4 \qquad\qquad R^{27} = R^3 + R + 1$$
$$R^5 = R^2 + 1 \qquad\qquad R^{26} = R^4 + R^2 + R + 1$$
$$R^6 = R^3 + R \qquad\qquad R^{25} = R^4 + R^3 + 1$$
$$R^7 = R^4 + R^2 \qquad\qquad R^{24} = R^4 + R^3 + R^2 + R$$
$$R^8 = R^3 + R^2 + 1 \qquad\qquad R^{23} = R^3 + R^2 + R + 1$$
$$R^9 = R^4 + R^3 + R \qquad\qquad R^{22} = R^4 + R^2 + 1$$
$$R^{10} = R^4 + 1 \qquad\qquad R^{21} = R^4 + R^3$$
$$R^{11} = R^2 + R + 1 \qquad\qquad R^{20} = R^3 + R^2$$
$$R^{12} = R^3 + R^2 + R \qquad\qquad R^{19} = R^2 + R$$
$$R^{13} = R^4 + R^3 + R^2 \qquad\qquad R^{18} = R + 1$$
$$R^{14} = R^4 + R^3 + R^2 + 1 \qquad\qquad R^{r7} = R^4 + R + 1$$
$$R^{15} = R^4 + R^3 + R^2 + R + 1 \qquad\qquad R^{16} = R^4 + R^3 + R + 1$$

If now the six transformations used in G.F.2^3 (table 10.2) are applied in turn to R, either separately or combined in succession, they produce exactly six of the elements just listed, namely:

Table A$_2$.2

R	R^4+R
$R^4+R^3+R^2$	$R+1$
R^4+R+1	$R^4+R^3+R^2+1$

For example $(1+1/t)*[(t+1)*R]=(1+1/t)*(R+1)$
$$=1+1/(R+1)$$
$$=1+(R^4+R^3+R^2), \text{ from table A}_2.1,$$
$$=R^4+R^3+R^2+1.$$

The fact that only six elements are obtained by this procedure suggests that the six transformations form a subgroup of the group G of order 30 whose elements we are trying to find. Also another transformation $1+1/t^2$ applied repeatedly generates fifteen of the 30 R-elements including the 3 on the left in table A$_2$.2 above.

Thus
$$(1+1/t^2)\times R=1+1/R^2$$
$$=1+(R^3+1)$$
$$=R^3,$$

and
$$(1+1/t^2)*R^3=1+1/R^6$$
$$=1+(R^4+R^3+1)$$
$$=R^4+R^3,$$

and so on. The fifteen transformations corresponding to these R elements may be found by repeated substitution of $(1+1/t^2)$ into itself.

$$(1+1/t^2)*(1+1/t^2)=1+1/(1+1/t^2)^2$$
$$=1+1/(1+1/t^4)$$
$$=1+t^4/(t^4+1)$$
$$=1/(t^4+1).$$

Substitution of this expression into $1+1/t^2$ leads to t^8 and so the left hand columns of table A$_2$.3 are built up, matching each transformation with its corresponding R-element. Sooner or later t^{32} appears among the transformations and can be replaced by t. This is because the R elements form a cyclic group of order 31 for multiplication so that t^{31} acting on any R-element whatever produces 1 as the result.

The fifteen transformations form a subgroup H of G. The next step is to find the transformations in the coset listed on the right of table A$_2$.3. We already know that they include $1/t$ among them and if we apply this transformation in turn to every element of H we obtain the full coset. In table A$_2$.3 each transformation is opposite its reciprocal, so that we can write down the corresponding R-element in either of two ways: (a) by using the table $A_2.1$ of reciprocals of R-elements, (b) by applying each transformation in turn to the element R.

Table $A_2.3$

G

	R-elements	Trans-formations		Trans-formations	R-elements
1	R	D	2	$1/D$	R^4+R
3	R^3	$1+1/D^2$	4	$1+1/(D^2+1)$	R^4+R^2+R
5	R^4+R^3	$1/(D^4+1)$	6	$1+D^4$	R^4+1
7	R^3+R^2+1	D^8	8	$1/D^8$	R^3+R^2+R+1
9	$R^4+R^3+R^2+R$	$1+1/D^{16}$	10	$1+1/(D^{16}+1)$	R^4+R^2
11	$R^4+R^3+R^2$	$1/(D+1)$	12	$1+D$	$R+1$
13	R^2	D^2	14	$1/D^2$	R^3+1
15	R^3+R	$1+1/D^4$	16	$1+1/(D^4+1)$	R^4+R^3+1
17	R^2+R+1	$1/(D^8+1)$	18	$1+D^8$	R^3+R^2
19	R^4+R^3+R+1	D^{16}	20	$1/D^{16}$	$R^4+R^3+R^2+R+1$
21	R^4+R+1	$1+1/D$	22	$1+1/(D+1)$	$R^4+R^3+R^2+1$
23	R^4+R^2+R+1	$1/(D^2+1)$	24	$1+D^2$	R^2+1
25	R^4	D^4	26	$1/D^4$	R^3+R+1
27	R^3+R^2+R	$1+1/D^8$	28	$1+1/(D^8+1)$	R^2+R
29	R^4+R^2+1	$1/(D^{16}+1)$	30	$1+D^{16}$	R^4+R^3+R

Combining the thirty transformations by succession then leads to table 11.3 based on the fifteen transformations 1, 3, 5, . . . 29 as a normal subgroup.

$A_2.2$ The group of transformations derived from G.F.3^3

The method used is much the same as for G.F.2^5. Table $A_2.4$ sets out the relevant powers of R obtained from an irreducible cubic equation $R^3=R+2$.

The first twelve powers run from 1–12 of the R-elements, then comes R^{13} which has the value 2 and is outside the group. The next twelve powers run from 24 to 13. For example the element numbered 23 is R^{15} which is the reciprocal of R^{11} as $R^{26}=1$. The corresponding transformations are derived from the subgroup listed in section 10.1.

$$t \qquad\qquad 2t+1$$
$$1+2/t \qquad\qquad 1/t$$
$$1/(2t+1) \qquad\qquad 1+1/(t+2)$$

The presence of modulus 3 in the field suggests that if $1+2/t$ is a transformation of the group we are trying to obtain, then also $1+1/t$

224

and $2+1/t$ are also in the group. Each such transformation can then be matched with its R-element by observing its effect on the initial element R.

$$(1+2/t)\ast R=1+2/R$$
$$=1+2(2R^2+1)$$
$$=R^2.$$

In this way a complete set of transformations can be found, one for each R-element.

The multiplication table for two transformations combined by succession is set out in table $A_2.5$ in which the elements 1, 6, 19, 12 are seen to be a normal subgroup of type D_2 (Klein's group).

Table $A_2.4$

G

	R-element	Trans- formations		Trans- formations	R-element
1	R	t	13	1/t	$2R^2+1$
2	R^2	$1+2/t$	14	$1+1/(t+2)$	$2R^2+2R+1$
3	$R+2$	$t+2$	15	$1/(t+2)$	$2R^2+2R$
4	R^2+2R	$2/(t+1)$	16	$2(t+1)$	$2R+2$
5	$2R^2+R+2$	$2+1/(t+1)$	17	$2+2/t$	R^2+1
6	R^2+R+1	$1+2/(t+2)$	18	$1+1/(t+1)$	$2R^2+R+1$
7	R^2+2R+2	$2+2/(t+1)$	19	$2+1/(t+2)$	$2R^2+2R+2$
8	$2R^2+2$	$1+1/t$	20	$1+2/(t+1)$	R^2+2R+1
9	$R+1$	$t+1$	21	$1/(t+1)$	$2R^2+R$
10	R^2+R	$1/(2t+1)$	22	$2t+1$	$2R+1$
11	R^2+R+2	$2+2/(t+2)$	23	$2+1/t$	$2R^2$
12	R^2+2	$2/t$	24	$2t.$	$2R$

(See table $A_2.5$ overleaf)

Table A$_2$.5

		A				B				C				P				Q				R		
		1	6	19	12	2	9	4	11	3	20	17	10	13	7	18	24	15	5	8	22	14	16	21	23
A	1	1	6	19	12	2	9	4	11	3	20	17	10	13	7	18	24	15	5	8	22	14	16	21	23
	6	6	1	12	19	9	2	11	4	20	3	10	17	7	13	24	18	5	15	22	8	16	14	23	21
	19	19	12	1	6	4	11	2	9	17	10	3	20	18	24	13	7	8	22	15	5	21	23	14	16
	12	12	19	6	1	11	4	9	2	10	17	20	3	24	18	7	13	22	8	5	15	23	21	16	14
B	2	2	4	11	9	10	20	3	17	6	12	19	1	22	5	15	8	16	23	21	14	13	18	24	7
	9	9	11	4	2	17	3	20	10	1	19	12	6	8	15	5	22	14	21	23	16	7	24	18	13
	4	4	2	9	11	20	10	17	3	12	6	1	19	5	22	8	15	23	16	14	21	18	13	7	24
	11	11	9	2	4	3	17	10	20	19	1	6	12	15	8	22	5	21	14	16	23	24	7	13	18
C	3	3	10	20	17	12	1	19	6	9	4	2	11	23	14	21	16	7	18	13	24	15	22	5	8
	20	20	17	3	10	19	6	12	1	2	11	9	4	21	16	23	14	13	24	7	18	5	8	15	22
	17	17	20	10	3	6	19	1	12	11	2	4	9	16	21	14	23	24	13	18	7	8	5	22	15
	10	10	3	17	20	1	12	6	19	4	9	11	2	14	23	16	21	18	7	24	13	22	15	8	5
P	13	13	18	7	24	14	21	16	23	15	8	5	22	1	19	6	12	3	17	20	10	2	4	9	11
	7	7	24	13	18	16	23	14	21	5	22	15	8	6	12	1	19	20	10	3	17	9	11	2	4
	18	18	13	24	7	21	14	23	16	8	15	22	5	19	1	12	6	17	3	10	20	4	2	11	9
	24	24	7	18	13	23	16	21	14	22	5	8	15	12	6	19	1	10	20	17	3	11	9	4	2
Q	15	15	22	8	5	24	13	7	18	21	16	14	23	11	2	9	4	19	6	1	12	3	10	17	20
	5	5	8	22	15	18	7	13	24	23	14	16	21	4	9	2	11	12	1	6	19	20	17	10	3
	8	8	5	15	22	7	18	24	13	14	23	21	16	9	4	11	2	1	12	19	6	17	20	3	10
	22	22	15	5	8	13	24	18	7	16	21	23	14	2	11	4	9	6	19	12	1	10	3	20	17
R	14	14	16	23	21	22	8	15	5	18	24	7	13	10	17	3	20	4	11	9	2	1	6	12	19
	16	16	14	21	23	8	22	5	15	24	18	13	7	17	10	20	3	11	4	2	9	6	1	19	12
	21	21	23	16	14	5	15	8	22	13	7	24	18	20	3	17	10	2	9	11	4	19	12	6	1
	23	23	21	14	16	15	5	22	8	7	13	18	24	3	20	10	17	9	2	4	11	12	19	1	6

APPENDIX 3

SENIOR WRANGLER PATIENCE

Readers who sometimes use 2 packs of patience-cards for recreation and are also interested in modular arithmetic may like to know the rules for Senior Wrangler Patience which is based on modulus 13.

Lay out 8 different cards, not kings, in a row. Call these the R-cards. Divide the rest of the pack into 8 piles of 12 cards and place one pile face upwards above each R-card. Cards visible at the top of the 8 piles are in play and may be used to build up 8 other piles below the R-cards. The lower piles are built up as follows:

(a) Double the R-card, e.g. if the R-card is 4 the first card required for the pile below it is 8; if the R-card is 9 the first card required, is 5 using 13 as a modulus.

(b) Add the R-card to the uppermost card in the lower pile to find the next card which is required.

R-card 4 produces the pile 8, 12, 3, 7, . . . ,

R-card 9 produces the pile 5, A, 10, 6, ...

(c) A pile is finished when it arrives at a king. If the patience is successfully completed all cards except the R-cards are in the lower piles with kings uppermost.

When all visible cards which fit into one or other of the lower piles have been played it is time for a new deal. The first pile from the left end of the upper row is dealt one at a time over the 8 places above the R-cards beginning with the next pile on the right. Successive deals are made later on from the second, third, fourth . . . piles and no new deals are permitted after the 8th pile has been dealt.

It is helpful to choose complementary pairs for the R-cards: a 4 should be balanced by a 9 as this helps to spread the requirements of the lower piles more evenly.

One merit of this patience is that one is never sure that one has got it out till the last few cards are played. Another is that it shuffles the pack which needs no further attention before being used for another patience!

227

APPENDIX 4

THE TWINS

There is an old song called 'The Twins' which my father used occasionally to sing at village concerts:

In form and feature, face and limb
I grew so like my brother
That folk got taking me for him
And each for one another.

It puzzled all our kith and kin;
It reached a fearful pitch;
For one of us was born a twin
And not a soul knew which!

One day to make the matter worse
Before our names were fixed
As we were being bathed by Nurse
We got completely mixed;

And so you see by Fate's decree,
Or rather Nurse's whim,
My brother John got christened me
And I got christened him!

This fatal likeness even dogged
My footsteps when at school;
And I was always getting flogged
For John turned out a fool.

I put this question fruitlessly
To everyone I knew
'What would you do if you were me
To prove that you were you?'

Our close resemblance turned the tide
Of my domestic life
For somehow my intended bride
Became my brother's wife.

And so year after year the same
Absurd mistakes went on;
And when I died the neighbours came
And buried Brother John.

                    Henry Sambrooke Leigh (1837–1883)

   The student's remarks, reported in section 6.13, in a discussion about
'brother of' as an equivalence relation, prompted me to look again at
the words of this song and to write an alternative last verse which
provides a quite different dénouement from that in the original.

'But though year after year the same
Absurd mistakes go on;
I've only got myself to blame
I am my brother John!'

BIBLIOGRAPHY

I. N. Herstein *Topics in Algebra* (1964)

W. Ledermann *Introduction to the Theory of Finite Groups* 5th ed. (1963)

D. E. Littlewood (a) *A University Algebra* (1950)
(b) *The Skeleton Key of Mathematics* (1949)

A. J. Moakes *The Core of Mathematics* (1964)

W. W. Sawyer *A Concrete Approach to Abstract Algebra* (1959)

H. S. M. Coxeter *An Introduction to Geometry* (1961)

N. N. Vorob'ev Trans. H. Moss *Fibonacci Numbers* (1961)

H. M. Cundy and A. P. Rollett *Mathematical Models* 2nd ed. (1962)

M. A. Jaswon *Mathematical Crystallography* (1965)

W. Wilson *Change Ringing* (1965)

A. F. Wells *The Third Dimension in Chemistry* (1956)

(Books relating to Chapter 5 are listed at the end of that Chapter.)

INDEX OF GROUPS AND RINGS MENTIONED IN THE TEXT

GROUPS

Type	Symbol	Page reference
Cyclic	C_2	1, 7, 37
	C_3	2–3, 10, 46, 145, 149
	C_4	7, 39
	C_5	3
	C_6	13, 23, 38, 39, 199
	C_7	164
	C_{10}	41, 142, 212–3
	C_{14}	151
	C_{15}	233
	C_{16}	43
	C_{18}	46, 150
	C_{26}	224
	C_{28}	151
	C_{31}	222
Dihedral	(Klein) D_2	19–21, 49, 61, 70, 158, 189
	D_3	24–27, 28, 29, 33, 34, 112
	D_4	47–51
	D_5	56
	D_6	36, 53
Quaternion		64
Direct Product	$C_2 \times C_3$	113
	$C_2 \times C_4$	62, 113, 121
	$C_2 \times C_2 \times C_2$	61, 115, 203
	$D_3 \times C_3$	116–120
	$D_3 \times C_5$	188, 222–4, 226
	$D_3 \times D_3 \times D_3 \times D_3$	211
Symmetric	S_3	181
	S_4	189, 191, 192
	S_5	

GROUPS—*continued*

Type	Symbol	Page reference
Alternating	A_4	122, 189
	A_5	51
Infinite	C_∞	31, 201
	D_∞	201
	$D_\infty \times D_1$	202
	$D_\infty \times D_1$	202

RINGS

Type	Symbol	Page reference
Integers (mod n)	Z_4	136, 137
	Z_6	16
	Z_8	135, 137
	Z_9	54
	Z_{10}	16, 54, 134
	Z_{12}	17, 41, 54
	Z_{15}	54, 145
	Z_{25}	151
Fibonacci		139–153

GENERAL INDEX

(Page references in bold type indicate a definition or explanation of a term or else the main discussion of a leading idea.)